PIMLICO

136

THE HIDDEN LANDSCAPE

Richard Fortey is a senior palaeontologist at the Natural History
Museum. He is the author of several books, including *Fossils, A Key to
the Past*.

THE HIDDEN LANDSCAPE

A Journey into the Geological Past

RICHARD FORTEY

PIMLICO

PIMLICO

An imprint of Random House
20 Vauxhall Bridge Road, London SW1V 2SA

Random House Australia (Pty) Ltd
20 Alfred Street, Milsons Point, Sydney
New South Wales 2061, Australia

Random House New Zealand Ltd
18 Poland Road, Glenfield
Auckland 10, New Zealand

Random House South Africa (Pty) Ltd
PO Box 337, Bergvlei, South Africa

Random House UK Ltd Reg. No. 954009

First published in Great Britain by Jonathan Cape 1993
Pimlico edition 1994, reprinted 1996

3 5 7 9 10 8 6 4

Printed and bound in Great Britain by
Butler and Tanner Ltd, Frome, Somerset

ISBN 0-7126-6040-2

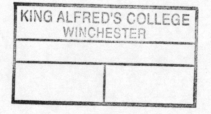

For Rebecca, with love

Contents

Picture Credits viii

List of Colour Illustrations ix

Preface 1

1 Journeys to the Past 3

2 Names, Origins, Maps and Time 15

3 The Oldest Rocks 29

4 The Great Divide 37

5 'Here be dragons': Caledonia 57

6 The Southern Uplands 74

7 The Land of the Ordovices 81

8 The Red and the Black 107

9 Fells and Dales 126

10 Coal and Grit 144

11 Lost in the Sands 160

12 Vales and Scarps; the Jurassic 175

13 The Weald 199

14 The Chalklands: Downs and Flints 215

15 The Chalklands: Beechwoods and Trout Streams 238

16 Tertiary Times 249

17 East Anglia: Sky and Ice 275

Glossary 295

Appendix 299

Acknowledgments 304

Select Bibliography 305

Index 307

Picture Credits

Many of the photographs have come from picture collections at the Natural History Museum (see author acknowledgements). The author and publishers also gratefully acknowledge the following sources for permission to reproduce the illustrations as indicated: *The Natural History Museum Picture Library, London*: Coral Thamnopora (col. plate), Oolitic limestone (col. plate), Nummulites gizehensis (col. plate), Fossil lycopod bark (col. plate); *The Director, British Geological Survey*: NERC copyright reserved: Smoo Cave (col. plate), Highland Boundary Fault (col. plate), Ongar Church (col. plate), Great Glen (p56), Conglomerates (p59), Cambrian Wall (p84), Slate (p86), Dartmouth and Kingswear Castles (p120), St Michael's Mount (p124), Thornton Force (p138), Grit Stacks (p157), Cresswell Crags (p164), Land's End (p173), Silbury Hill (p236), Dry Valley (p247). The endpapers show the BGS Geological map of Britain and Ireland, NERC copyright; *Aerofilms Ltd*: St David's (p11), Solway Firth (col. plate), Skiddaw (col. plate), Culzean Castle (p106); *Norfolk Museums Service* (Norwich Castle Museum): *Mousehold Heath*, etching by John Crome, c.1810-11 (p285); *Syndication International*: Rye (p203), Bath (col. plate), Dartmoor (p171); Holy Island (p143), Stonehenge (p266); *S & O Mathews*: Pargetting (p226); *J. Fortey*: Round Towered Church (p226); *S. Godwin*: Ashmolean Museum (p195), Hertfordshire Puddingstone (p265).

List of Colour Illustrations

Between pages 86 and 87

Smoo Cave, near Durness, Sutherland

The Moine Thrust at Knockan Crag

The Solway Firth

The Highland Boundary Fault

Plants that flourish on the peaty soils of ancient Caledonia: Club moss, *Drosera*,
    and *Silene acaulis*

Dob's Linn, Moffat, Southern Uplands

Cader Idris, a mountain largely composed of Ordovician volcanics

Skiddaw from the air

Old Red Sandstone: St Anne's Head, Pembroke, South Wales

The Rhynie Chert

Thamnopora, imprinted in limestone

Arthur's Seat, Edinburgh

Lepidodendron, an important contributor to coal

Bloody cranesbill, a limestone lover

Between pages 214 and 215

The Bog Asphodel, a plant of acid bogs

'New Red Sandstone', found in South-west England

Oolite limestone

Bath: the definitive stone city

Chesil Beach, viewed from Portland Bill

Stairhole Cove, Lulworth

Sunken lane in the Weald

Geology and scenery, near Box Hill, Surrey

Red Chalk at Hunstanton, Norfolk

Plants on Chalk: Bee orchid, Wild thyme, and *Morchella esculenta*

Alum Bay, Isle of Wight

The wild Gladiolus, *Gladiolus illyricus*

Nummulites gizehensis

Greenstead Church, Ongar, Essex

Cottages in Norfolk

Fingal's Cave, Isle of Staffa, West of Mull

Preface

THE RECENT popularisation of science is part of a modern passion for explanation. Most books in this genre are concerned with exposing the mechanics of the space/time continuum, or the workings of the body, by stripping down the complexity of the world to its components. The intention is to show that, deep down, everything is readily comprehensible. At the end one is much the wiser, but as when a complex conjuring trick is explained as a mere mechanical contrivance there lingers a vague feeling of disappointment. Somehow, the enjoyment of the trick is more satisfying than the explanation.

When Heather Godwin asked me to write a book with the geology of Britain as its theme I decided to avoid the purely analytical path. Instead, what I wish to explore are the *connections* between geology and landscape – the deeper, geological reality that underpins natural history, and even our own activities. I want to celebrate the richness of the geological history of our islands, and to show how events that happened hundreds of millions of years ago still control the lie of the land and the plants that grow upon it. The intention is to enrich the reader's awareness of our extraordinary past.

This intention makes my geology a more personal account. I will explore curious byways of the landscape as much as narrating its geological history. Jacquetta Hawkes' book, *A Land*, was published nearly forty years ago. It was one of the books that inspired me to take up science; the facts on which that book was based are now somewhat out of date, but the inspiration remains. If I can stimulate a sense of wonder in a similar way, this book will have succeeded.

There has to be an historical narrative, which, as it happens, runs broadly in tandem with the geography of Britain. Lying beneath the thin skin of recorded history in our islands, geology has the same role in landscape as does the unconscious mind in psychology: ubiquitous but concealed. This is the hidden landscape.

1

Journeys to the Past

TWENTY YEARS ago I travelled to Haverfordwest to get to the past. From Paddington Station a Great Western locomotive took me on a journey westwards from London further and further back into geological time, from the age of mammals to the age of trilobites. The train soon quit the flat Thames Valley beyond Reading, and with it the soft sands and hard cobbles of the Reading Beds, laid down when there were mammals and birds on land, and crabs in the sea, and when the world would have felt familiar. Then I was speeding over the Chalk, and back in time to the company of dinosaurs. The train travelled on towards Bath, where Jurassic limestones and shales take turns across the countryside, the former proud with ancient corals, the latter dark and low, with ichthyosaurs and plesiosaurs, 'sea lizards' that grasped Jurassic fish and ammonites. Of ammonites, plesiosaurs and ichthyosaurs, nothing remains but their fossils. Under the River Severn and into Wales, and I was back *before* the time of the dinosaurs, to a time when Wales steamed and sweated, not with the fires of smelting, but with the humid heat of moss-laden and boggy forests in coal-swamps, where dragonflies the size of hawks flitted in the mist; and then on back still further in time, so far back that life had not yet slithered or crawled upon the land from its aqueous nursery.

At last I arrived at Haverfordwest — I had come back far enough for the rocks to yield trilobites. Trilobite land lay all around; to the west as far as the sea at St David's, imbued with the mystery of the origin of all things. First I walked along the valley of the River Cleddau where the Silurian rocks could be found. I walked there from the station. The route went past the gasworks and out on to the river bank where a crude towpath was a regular dog walk for the local people.

Giant dragonflies flitted through the luxuriant silence of the Coal Measure swamps.

Rock peeped out from behind bramble bushes, or in little scrapings by the path. Four hundred million years ago these rocks had been soft silts and muds on the floor of the Silurian sea, and now, hardened by time and elevated by tectonics, they revealed their proper order along the Cleddau Valley. I had nothing with me except what I could carry on my back, and the most important item of all was the geological hammer. These soft rocks were not difficult to break.

I pulled out a substantial lump of the crude rock, and then, holding it to the light like a connoisseur examining a rare vase, I picked out a line marking the direction of the former sea floor. Only one plane through the rock would correspond exactly with the fossilised sea bed; this is where the fossils lie. Break the rock in another direction and I would risk destroying the fossils. For fossils lie on the *bedding* – and what an appropriate word for tucking up past life into an endless sleep in the rocks.

Most geological hammers have a square, bashing side, and a chisel side. A gentle tap with the latter is often all that is needed to break open a bedding plane, and awake the sleeper. So tapping and tapping at the Cleddau rocks I obtained a good break along the bedding; every time such a crack opened up, the two sides of the rock were parted impatiently: inside could be who knows what treasure. When nothing was revealed, and the rock split into two to show only the grey, speckled face of a barren Silurian sea floor, I plucked a new piece of rock from the rock face, and split it. And then, suddenly, a trilobite! No longer than my thumb, but crouched in the rock as if waiting to be released from its sleep in stone, I cried a shout of pure joy at the discovery. I can still remember my anticipation as the rock piece parted comfortably in my hands, almost as one might cut a deck of cards: no strain or grinding. Then there was astonishment in suddenly finding this complex creature, so perfect though so old, with a pair of petrified eyes visible to the naked eye even in the misty light of a Welsh afternoon, eyes that had last seen the world more than four hundred million years ago – before the first, humble liverwort had colonised the dampest shore, before sharks, before flies, and how unimaginably far before humankind.

The trilobite had the shape and the feel of an artefact; something of the neatness and symmetry of a medallion. Like a medallion it could sit comfortably in the palm of my hand. The fossil showed a head, with its eyes, and a middle lobe, a tail, and a thorax with perhaps a dozen segments – a complicated animal despite its antiquity. I remember a curious feeling, as if in some way this revelation to my hammer after so long asleep in the bedding of the rock had not just been a matter of serendipity. Perhaps I had been intended to find that trilobite, to make the blow upon just that piece of rock, and to release that very messenger from the past into the world to tell its story. I became aware of the continuity of things. There was a thread running between this trilobite and this investigator. At the time the only feeling I would have been able to articulate was one of *specialness* of the moment and of the place, a kind of contentment I could hug to myself. The excitement of the find was physical, like any kind of hunting. But the metaphysical component was there, too, at the very least a species of shock to be made so aware of how long this place had existed as a haven for life – why else should this stone-bug, preserved in fossil clay in part and counterpart, have seemed as if sent to me as a talisman?

I knew enough then to avoid trying to clean up my precious trilobite in the field. Tissue paper was kept for very special discoveries. The more routine finds were wrapped in newspaper. In this part of Wales there was no youth hostel and I had to treat myself occasionally to a cheap bed-and-breakfast. In the evening, under the inadequate illumination of a 40-watt light bulb, I took out the trilobite again and gloated.

Welsh bed-and-breakfast at the cheapest end of the market tended to be strange and dark, mostly cold rooms at the back of inns with few customers. They do not have the jolly hanging signs of English pubs, but a stark name painted on the wall – something simple like THE NEW INN. A friend dubbed one cheerless example The Misery Arms; by an odd perversion of the normal function of the inn this particular place was dedicated to dispensing gloom rather than cheer. A request for a half of bitter was greeted by a deep sigh, and 'I shouldn't wonder if it's flat' or 'very thin, our bitter'. Sandwiches were unavailable, the crisps as flexible as fallen leaves. I have often wondered about this substitution of gloom for what should be the jolliness of a good inn, and it seems that the only explanation lies in the ineluctable association of having a good time with guilt in the non-conformist tradition. There is never anybody behind the bar when you arrive. You stand there shifting from foot to foot for a while, hoping somebody will appear. Nobody does. You try a variety of throat-clearings, coughs and gargles. Finally, you yell 'I say! Hello!' or something equally daft, and eventually your barman appears. 'I've been out the back . . .', he says, unnecessarily. I suspect that he knows you are there all the time, and is allowing sufficient angst to be built up, to expiate the guilt of the half of bitter you may, if you are lucky, eventually get.

For me, at that time, these curious discomforts had no importance beside the excitement of discovery. Imagine the odds against my having found this particular trilobite. This animal, like all organisms at this early part of the Silurian period, lived in the sea. A deep sea covered most of Wales then, but only in some parts of this sea were conditions suitable for trilobites to thrive. Other parts were stagnant – starved of the oxygen all animals, even trilobites, need to breathe and grow. Add to this the fact that trilobites had enemies in the Silurian sea; we can find the occasional specimen with healed bite marks. These were the ones that got away; many more never had the chance to leave their remains behind. And what if my trilobite, having died, were discovered by creatures scavenging for food? These would soon have dismembered the carcass, scattering pieces of my animal here and there over the sediment surface so that it could never be reconstructed. Most marine life does get broken and redistributed in this way, and this process keeps the nutrient cycle turning over. So perhaps the trilobite died under unusual circumstances, ones that favoured its preservation. Perhaps it died under a blanket of sediment generated during an exceptional storm? Murdered and buried by

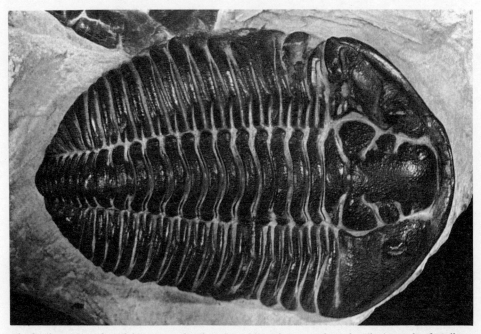

The fossil carapace of a trilobite, 'so perfect though so old': Calymene, *from the Silurian rocks of Dudley, Worcestershire.*

a single act. But then how often would such storms occur? Maybe only one a century would have violence enough. Remember our sense of astonishment when trees – oaks even – which had stood inviolate for three hundred years or more were devastated by a single night's storm in October, 1987? Possibly the events needed to entomb the trilobite were as infrequent, so that it became the chance casualty of an exceptional event. The sediment itself must have been of the right kind to preserve the fossil. The trilobite in my hand was not complete: in life, it would have had a battery of little legs, like those of a prawn, not a trace of which remained. Those legs lacked the hard shell usually necessary to become preserved as fossil material; only the hard carapace was mineralised and hence fossilised. But even such resistant material can be dissolved and leached away in the wrong kind of rock. My mudstone behind the gasworks in Haverfordwest was of the right type to cradle safely the carapace of the trilobite for its long entombment.

But this is nowhere near the end of the story; much can happen to the rocks themselves. Time piles rock upon rock. The sea comes and goes with the passing geological ages. Unless other events intervened, my trilobite would have become interred within an ever deeper pile of sediments, to a depth possibly beyond the deepest coal seam, and buried into an obscurity from which it would never emerge. But often in geology that which is buried is destined to rise. Phases of mountain building throw up rocks that were once deep beneath the surface. The British Isles have been through no fewer than three such phases since my trilobite scuttled about on the sea floor. The first of these – the great Caledonian convulsion – was responsible for disinterring my fossil. But even then it must have had a lucky escape; these times of orogeny also destroy countless fossils. In some cases fossils become twisted and distorted beyond recognition, in other cases they may be baked in the furnace of the interior of the Earth before they have a chance to rise to the surface. Rocks are continually recycled. Sediments become sedimentary rocks, sedimentary rocks are entombed deep in the crust, and these in turn may be squeezed and distilled back into hard metamorphic and even igneous rocks. In this great cycle the evidence of life may be destroyed in a moment. There are areas of the world the size of England – in parts of the Andes, for example – where the former sedimentary rocks have been so maltreated that hardly a fossil survives. The trilobite escaped all this. But even when lifted to the surface, out of reach of the transforming fires of the interior of the earth, there were other obstacles in the way of its discovery. It could have been reburied beneath younger sediments: indeed, it almost certainly was, and these sediments have been stripped away in turn by erosion. The last ice age (geologically speaking, an event of yesterday, because the ice only retreated seven thousand years ago) increased the rate of erosion enormously and the fossil might have succumbed to washing, or grinding, or freezing. And finally there is chance that makes a scrape in the ground here rather than there; the decision of who-knows-what town councillor of Haverfordwest to widen the path, thereby exposing a little cliff of rock to my hammer. And finally, and most importantly, my hammer fell on the right spot, the rock cleaved in the right way, and I did not fumble and lose the specimen in my excitement.

So my discovery was a concatenation of chances, linking across hundreds of millions of years, serendipity a thousand-fold. Could such a chain of circumstances still be chance alone? Or perhaps my awareness of the complexity of the chain invested the find with mystery? Imagine a kind of relay race, in which the trilobite is the baton. Time runs the baton through hazard after hazard, but at each test the token survives, to be thrown on to another lap of chance and survival.

We cannot conceive of the sheer expanse of time. We can count tens readily enough, and have an adequate notion of what a few hundreds might look like. Even trying to think of thousands causes an approximation in our conception – could we really tell whether a heap contains a thousand pebbles, or three thousand? There is a game at every village fete which relies upon the ambiguity of numbers of this order. Usually it is something like 'Guess the number of sweets in the jar'. The closest guess collects the prize. But glance down the list of guesses and you can see how approximate is our vision of numbers even in their hundreds – there may be a threefold difference in the estimate of the number of sweets in a jar that can easily be held in both hands. If we become confused with mere thousands, how do we appreciate a million, let alone ten million, or a hundred million? Yet palaeontologists will toss around their millions of years with the ease that other people talk about last Friday. My vision of the trilobite fossil as a baton passed down through the geological ages is actually absurdly speeded-up. The mountain ranges come and go with the ease that one can push up a fold in the table cloth. As each geological cycle is completed my hand-sized token is flipped nonchalantly on to the next circuit. Yet perhaps the Keystone Cops version *does* serve to animate the inexorable slowness of real time. It reduces it to the scale where we can view events through the screen of our own mere three or four score years. So our human chronology domesticates time.

My trilobite was invested with the mystery of time – it was the baton handed to me, at last, from the Silurian. Even if I could only understand time as if in a cartoon I could picture the extraordinary chain that linked me by way of my talisman with the period before the first four-legged creatures. Surely such a messenger should bring momentous news: the arrival of the messenger was itself momentous! It came as a spyglass to see back into the past, as a means to see this little piece of Wales near the gasworks in Haverfordwest as it was 400 million years ago. Such excitement made my uncompromising bed in the Welsh bed-and-breakfast seem a small concern.

It was further to the west, near St David's, that I began to appreciate more the links between the rocks beneath and the land above, and between our own lives and distant prehistory. St David's is a little town set at the western tip of South Wales, and it is not much bigger now than it was two hundred years ago. It used to be the cathedral city of Pembrokeshire, and is now part of the larger county of Dyfed, a confection produced by the reorganisation of the Welsh counties in 1972. It is an inappropriate confection, too, because Dyfed now combines the fiercely nationalistic hills near Carmarthen, in Welsh Caerffyrddin – the Castle of Merlin – with the anglophone, seabound, little-England-beyond-Wales that runs from Pembroke to St David's: two counties if ever there were such. St David's cathedral eschews the hard look of the Welsh chapel. It nestles in a hollow away from the gales of winter, and is tiny for the seat of a see, but with all the dignity. The ruined bishop's palace nearby asserts the Church of England. The whole of St David's has the feel of a market town from Gloucestershire cleverly floated off and moored off the Welsh coast. The cathedral is constructed from the rock of the place, much of it wonderfully mottled with lichens. Purplish sandstones, and other hard building stones have ensured its endurance. Lichens love to grow where there is much moisture and little pollution. They will grow *anywhere*, for of all organisms they demand the least, living off the very rain itself, with vitamin supplements supplied from the stone over which they grow. This lack of demand can be their undoing, for where there is pollution in the atmosphere the lichens gulp in lead along with their nutriment. Eventually they die. A few hardy species, such as those on the tombstones in Highgate Cemetery in London, are all that survive pollution. But in St David's the moist air blows in clean from the Atlantic ocean, and many species of lichens compete to soften any freshly hewn stone. They dapple, they paint. And this is one reason why the cathedral there seems to sit so comfortably in its own place, as much a growth from the land itself as an artefact. Beyond the cathedral the open countryside shows the occasional outcrop of those rocks from which the cathedral was built. The building is truly part of the landscape. Such trimmings as there are inside the building from 'foreign' parts, such as the Purbeck 'marble', are like light touches of cosmetic which serve to heighten its native beauty.

But to see the rocks themselves you must go to the sea. The cliffs near St David's are now famous. Much of the cliff path is owned by the National Trust, and it winds up and down all around the old county of Pembrokeshire. You can walk to Fishguard if you have a mind. And everywhere there is rock. Rock plunges vertically into the sea where the breakers chew at spiky slabs two hundred feet below. In other places there are folds; the bedding is twisted, and convoluted, and thrown into crazy spasms as if the strata had been in the hands of a demented pasta spinner. Then, suddenly, along a vertical crack, the rocks will change: black shale will give way to purple sandstone, or silvery slate to red shale. These cracks are just what they seem – breaks in the fabric of the

The cathedral at St David's, Dyfed, a building at ease with its geological foundations.

Earth. Quiescent now, movements along these faults, between the rock bodies on either side, once generated earthquakes as powerful as that which demolished San Francisco. Below, the sea, ever searching out the weak spot, has picked out the fault to make a cave. Hardly any of the rocks here seem to be horizontal, as they were when they were laid down beneath the sea. On these cliffs, we can feel the strength of the Caledonian convulsions twisting and breaking the rocks, conjoining what was once separate in time and space, separating what was once juxtaposed. This is a scrambled piece of country.

We can see the real complexity of the rocks. Near Haverfordwest much was hidden beneath trees and town and field – but here everything is on display. How different these buckled cliffs are from the stepped and ordered strata of

11

other parts of Britain: the banded Lias of Dorset, the chalky Seven Sisters.

Near St David's, the little fishing villages of Solva and Abercastle crouch close to the cliffs and almost seem a part of them. Just like the cathedral, the cottages grow from the landscape, for all that they are now smart homes rather than the houses of poor fishermen. The connection I felt with the remote past when I cradled that trilobite in my hand applies no less to the whole human landscape.

When I first caught sight of the St David's peninsula from the bus at Newgale I knew that here was a special and ancient place. For most of the rocks are Cambrian, the very oldest rocks with fossil shells. Cambria is a Roman word for Wales, and so it is Wales that donated its name to a geological system that is now known around the world. Near St David's, the largest British trilobite can be found, *Paradoxides davidis*. The largest specimen of this species is at the Natural History Museum – it is sixty centimetres long, a trilobitic monster. *Paradoxides* was once the villain in a rather second-rate science-fiction novel entitled *The Night of the Trilobites* in which the trilobites stirred from the rocks, came alive and sought to wreak havoc on the present day inhabitants of their ancient homeland. But these Cambrian fossils have suffered with the rocks that contain them: cracks dissect the exoskeleton, and they have been as twisted as the rock on which they lie. Even the Haverfordwest trilobite was a more promising candidate for revivification. But the very antiquity of *Paradoxides* – just the fact of its suffering in the rock during the 500 million years since it was alive – gives it a particular excitement. Its battered appearance testifies to that chain of special circumstances which brought it from the muddy sea floor of the Cambrian to the steep cliffs at St David's. Such a giant was as nothing compared to the forces within the earth that might have destroyed it. It seems to say: I have survived, but how narrowly . . .

There was a time when one could examine a whole specimen of *Paradoxides* in a cliff in a small bay called Porth-y-Rhaw. Shot through with cracks, this giant was uncollectable. The animal which once crawled about the Cambrian sea floor had been elevated into a vertical position along with the surrounding strata. The soft mud which had once been its home had been changed to a hard, splintery mudstone. Maybe the spray from a winter storm would once again anoint the trilobite with sea water for the first time for hundreds of millions of years. The trilobite was there when I first visited the spot, but when I returned twenty years later I found it had gone. Someone had tried to collect it from the cliff: only crude gouges remained. The excavation was defeated by the splintery rock and the dense pattern of joints, and the trilobite had simply crumbled into meaningless shards. Thus one of the most tangible links into the Cambrian I have ever seen was destroyed by an hour or two of thoughtlessness. The baton which had passed through the Caledonian crisis, and had outlasted the Ice Age, was thrown away at last. Worse than ignored, this message from the past had been destroyed.

A similar disregard for the past, and ignorance of context, has led to the erection of a brightly-coloured service station just outside St David's. This flimsy, flat-roofed, synthetic thing is like an encampment by an invader. Far from settling into the land it *announces* itself; it is both temporary and intrusive. It has nothing to do with the land. It has nothing to do with geology. It ignores the past. Nor does it provide a substrate for a clothing of lichens which might soften its geometry. Compare the cathedral, which seems to belong to the St David's peninsula, to grow from the geology, and to be part of that fateful continuum running from the Cambrian to the present. This petrol station echoes the hack marks surrounding *Paradoxides*. The connections with the past have been severed, the continuity broken. It is not just that modern construction demands that red brick push its uniformity into every corner of the land, even though it sits uncomfortably where there should be rock or stone. Rather, the character of the land is being obscured, its diversity diminished in a thousand ways. Much of the character of our country is governed by its geology, and determined by the rocks. The service station demonstrates how this is changing, as the quick commercial fix comes to dominate the local and the contingent. The beauty of Britain resides in the contrasts between its regions, and in the multiplicity of these regions. Much of this depends on geology. Those complex and chancy links into the past affect us even now – and if we lose awareness of them, that loss diminishes us. Perhaps this was the message that the trilobite brought me in Haverfordwest.

The geological map of Britain is a blotchy and complex affair like the battered complexion of an old drunk. It is a small area: in some parts of the world an area as large as Britain might be coloured with a single shade, underlain by one monotonous rock type. It is much-in-little. This diversity has given us our scenery, our older buildings, our natural vegetation, something of our history; in short, our character.

The covering of soil which embraces the human part of the history of our islands is but a thin one; vegetation and buildings are a skin covering the geological patchwork. I envisage rolling back this skin, as one might a carpet, to reveal the rocks beneath. In some places the carpet would be vanishingly thin, as over much of Wales and Scotland, in others comparatively rich and deep, as in the Severn Valley, or over much of the agricultural land of the south-east. When the covering is removed the geological map becomes explicit. How easy it is then to appreciate the way in which our islands have been stitched together from a hundred different rock formations. How simple it is to see that the younger formations lie upon the older as blankets lie one on top of the other on an undisturbed bed, or that in other regions this order is disarranged as if the blankets had been grasped by a giant hand, and twisted and crumpled this way and that, or even overturned completely. How clearly we can see that the great masses of granite that make up Dartmoor or the Cairngorms cut through the surrounding rocks like great boils welling up from below, or that some parts of the country are folded into gentle bowls, or puckered into massive domes. This is the underlying reality of the character of the land. Those parts of the country seeming at ease or in harmony still reflect this fundamental map. The flowers that grow wild in the hedgerows acknowledge this foundation. The fungi know it. Farmers fight it or celebrate it. Houses once grew from it.

This is a book about connections between geology, natural history, and ourselves. The intention is to inspire a way of looking at the landscape, rather than to enumerate its thousand quirks or to act as gazetteer. The basis in geology is, naturally, a scientific one, and where this is unfamiliar I shall try to explain it. The architecture, or flora, or scenery that grow from it will likely be better known. The progression in the book is from ancient to modern, through geological time. This is also a journey, very broadly, from North-west Scotland to South-east England. It will be a journey full of diversions, digressions and distractions.

2

Names, Origins, Maps and Time

MAN IS a taxonomic animal.

Taxonomy is the discipline of naming plants and animals, but in a wider sense it describes the giving of names as a tool for understanding the world. Science, and especially a descriptive science like geology, revels in a welter of names. 'Blinding with science' describes being dazzled and confused by words and terminology. One of the first characteristics of pseudo-science is the invention of an arcane vocabulary which both binds together the fraternity, gives a spurious objectivity, and excludes strangers. This partiality for names excites suspicion: surely honest folk call spades spades.

Such suspicion is a good thing, but taxonomy remains central to our understanding of the natural world. This is not the dubbing of things for the sake of it; a name can be shorthand for understanding. The more precisely you are able to name a flower, the more you know about flowers. Later, I will explain how the naming of the great Moine Thrust in the North-west Highlands served to unscramble the history of thousands of square miles of the British Isles. Any visitor to the Highlands can now talk about the Moine Thrust with an easy familiarity, but just by using the word 'thrust' he acknowledges the resolution of a century of scientific squabbling about the structure of Scotland. So names label concepts that may have had a long struggle to be born. To employ them is to give homage to those who have made the discoveries that have changed the way we see the landscape.

John Fowles, in his book *The Tree*, has pointed to an obsession with names – or naming – which may interfere with our response to the thing named. This is something like the mania that can take over in an art gallery, when, rather than look at the pictures, we shuffle about looking at the labels in the curious

15

belief that this will somehow help us to understand the artist. Or it may be a reference to the extreme form of 'twitchery', whereby bird enthusiasts pursue some rare warbler or bunting, with apparently no end in view but to tick it off a list of British birds seen in a lifetime. This is even more bizarre when the bird in question is a straggler, far from its distant home where it is as common as a lark. The aim here is the list, the compilation, not a greater understanding or appreciation of the object.

Train-number collecting of this kind is usually harmless, but to lumber the name-giving process as an obsession only with lists is to miss the important point. Human cognition often begins with labelling, with defining, with naming. Children advance from naming 'birds' as one division of 'animals' to separating pigeons from crows and sparrows; those who progress further to name all the British species can only do so by learning to observe closely, to discriminate differences in plumage, to appreciate birdsong, and to recognise habits. This is not just ticking off items in a list, it is learning to appreciate shades of meaning which cannot be recognised without depth of understanding. This is almost a definition of taxomony. Nor does it follow that aesthetic appreciation is somehow blunted by this taxonomic knowledge. There may be those to whom getting the name right is more important than enjoying the thing itself, but there must be many more for whom the discrimination of species opens the way to appreciating the marvellous richness of the world. The opposite extreme would be a taxonomic ignoramus who is somehow still capable of responding to nature and who would derive the same pleasure from a spruce monoculture as from a rich, mixed deciduous woodland simply because he could not tell the difference. Presumably he would soon succumb to belladonna poisoning because he would be unable to tell deadly nightshade from blackberries.

So names *matter*. W.H. Auden, whose brother was a distinguished geologist, had an early interest in geology, and knew the feel of names in his early days: 'A word like *pyrites* was for me not simply an indicative sign, it was the Proper Name of a Sacred Being.' Some peoples have names of such dread significance they dare not be spoken aloud. The Inuit (Eskimos) have a vast list of names for different colours and conditions of what to us would be simply 'white snow'. Names describe what is significant in the world, and this can vary from culture to culture. Scientific names attempt to provide a common language, and so describe what *is* in the world, transcending cultural differences. To name 'thrust' or '*pyrites*' is partly to acknowledge an awareness of what they mean, and this meaning may not have been obvious once. At a practical level the scientific names of plants, animals and fossils are used the world over in the same way. The trilobite *Calymene* will be known to a Chinese by the same scientific name whatever its local vernacular name.

So it is impossible to dodge using specialised names. They provide the vocabulary of history. There is nothing intimidating about names. Small children soon master *Tyrannosaurus*, which is a technical name; indeed, they seem to relish such labels in all their Graeco-Roman, polysyllabic splendour. Some even manage *Pachycephalosaurus*.

The first names to introduce are those of the main classes of rocks: igneous, sedimentary, metamorphic. These three names encapsulate some of the great struggles in geological science. It has not always been clear how these rocks were formed, nor even that they represented rocks of a fundamentally different cast. Igneous rocks originated from molten magma deep within the Earth's crust, which solidified as it cooled. They are nearly always crystalline, and frequently very hard. Granite, basalt, pumice, and gabbro are all species of igneous rocks. Because of their hardness and resistance to weathering igneous rocks may form the major features in a landscape: Arthur's Seat in Edinburgh, The Black Cuillins in Skye, Lundy Island, Bodmin Moor, Strumble Head on the coast in western Wales. They protrude against sky and sea. They can be attacked by the weather (as can everything except gold and platinum), but this attack is slow, extraordinarily slow. This is why Lundy Island remains, battered by every storm, while all the rocks around it have been eroded into oblivion.

The two main types of igneous rocks are volcanic, the product of eruption, and plutonic, which formed deep in the Earth's crust in the very bowels of the Earth. Those formed in Pluto's domain often cooled slowly, and had time to grow large crystals. Look at almost any city kerbstone, which will likely be made of granite, and the big pink or white crystals can be seen with the naked eye. Plutonic rocks may form not merely mountains but mountain ranges. Vulcan's rocks were formed violently, from a forge of fire, spewed from vents or extruded in incandescent clouds, and forged in different ways, too, so that they can look quite different: twisted or blocky or piled in flow upon flow. They can form fast, spread wide and cause much damage – as did the Mount St Helen's eruption of 1987 – or they can flow like burning treacle and be easily sidestepped. Such flows are a part of everyday life in Hawaii. Yams can even be cooked in the hot ashes.

Sedimentary rocks are nothing more than ancient sediments, and most of these accumulated slowly, a few millimetres a year. Most rocks of this kind were formed beneath the sea. This geological fact is one of the few that has penetrated common consciousness. On one field trip I met a farm worker who, on hearing that I was a geologist said: 'They say that it was all underneath the sea round here at one time.' This was accompanied by a vague gesture at the landscape and a look of sly scepticism. I didn't dare reply that the whole scene, enveloped in drizzle, had been beneath the sea, not once, but several times, and not so long ago, geologically speaking, had been subtropical. One can imagine the laughter at The Grubbing Mattock Inn later that day.

The former deposits of lakes and rivers are also sedimentary rocks; these are where one can expect to find fossils of plants or quadrupeds. Even desert sands may be preserved as a very particular kind of sandstone. What most sedimentary rocks have, which other kinds of rocks do not, is evidence of *bedding* – stratification showing where the former sea, lake, or river bed once was. It is bedding which defines the layers in the Lias, which sets the profile of the steps on the limestone hillsides in Derbyshire, and which I saw contorted by paroxysms of the Earth in the cliffs in Pembrokeshire. Sedimentary rocks today are being laid down over much of the Earth. Imagine a satellite picture of the Earth: all the blue oceans covering much of the surface of the globe are where sediments accumulate. You may be able to pick out a discoloration where one of the great rivers, the Amazon or the Nile or the Ganges, spills into the sea: this stain is the sediment itself. The rivers have flood plains floored by more sediment. And lakes like Baikal or Victoria or Ontario are floored by muds that will one day become rocks. You may even be able to pick out the desert areas on which sand dunes migrate: some of these, too, will eventually become rock. But equally there are other areas of the world, and not just high mountains, where the ground wears away, eventually, and these areas will leave no trace in the rocks. Because fossils are found in sedimentary rocks, it follows that there will be certain kinds of animals and plants which live in erosional sites – mountain lions, or alpines – which are destined to leave no fossil record. If they become extinct it will be as if they had never been.

The third great class of rock is metamorphic. Metamorphic rocks can be generated from either igneous or sedimentary rocks, so they are, in a sense, secondary. Metamorphism is change, and in rocks this means the transformation of limestone into marble, granite into granulite or shale into slate and schist. Heat

Sediments accumulate in the world's oceans, and even in deserts (Africa and the Arabian Peninsula, NASA)

and pressure are the agents of change, and both sedimentary and igneous rocks can be variously squeezed and heated to take on a new character, and acquire a new name in the process. This pressure and roasting almost invariably goes on deep in the Earth, driven by the inexorable processes of geological change: the movement of plates, heat from magma. Thus it is that metamorphic rocks are usually found at the surface where mountains are, or have once been, because many mountains are the legacy of the turbulence of the Earth. Metamorphism took place in the very roots of a mountain chain. 'Changed' rocks present many appearances, because metamorphism itself operates in several ways, and the starting materials are legion. Glistening crystals of mica are often present; in the Highlands of Scotland they catch glancing sunbeams or flash in car headlights. How many times must such a source have been traced in the expectation of finding a diamond – only to discover a micaceous mineralogical mirror. Many metamorphic rocks are banded or striped with layers of different texture which are contorted and thrown into ripples or ruckles or fantastic waves. These layers do not necessarily parallel one another for long distances (as the bedding generally does in sedimentary rocks) for they are the expression of pressure and heat and may veer this way and that on the whim of who knows what geological enormity forged beneath the weight of an ancient Himalaya.

Equipped with these three words, we can read the first and simplest geological map of our islands: this shows the distribution of igneous, metamorphic and sedimentary rocks.

This first map tells us important things. Sedimentary rocks dominate alone in most of England, with igneous rocks in Wales. Sedimentary rocks produce arable England, farming England, where soil expresses the character of the underlying sedimentary rocks. Rock types dictate whether cows can graze, or whether cereal or beet is grown. Resistant rocks, weathering slowly, produce uplands like the Pennines, where sheep prosper almost alone. Clay vales produce deep, rich soils. Sandy soils suit carrots; thin, chalky ones barley. The character of the land follows the outcrop of the rock types in broad swathes or strips miles wide. The rocks dictate whether wattle and daub, or stone, or flint rubble was used in vernacular architecture. They dictate whether old man's beard, heather, or devil's bit scabious grow by the wayside. In subsequent chapters these interconnections are the subject of detailed exploration. The metamorphic rocks are dominant in the Highlands. This tells us immediately that this part of the country must have been part of a mountain chain at one time – but when? And why does so little of it remain? Here, too, the vegetation is characteristic: peaty moorlands, now partly disappearing beneath swathes of planted conifers; Scots pines; old oak woodland thick with lichens. There is little agriculture now – except in the North-east – and this is also the exception geologically, as it is underlain by sedimentary rocks rather than metamorphic. Most recently, ice has scoured this land, leaving its legacy in lochs and perched boulders. As for igneous rocks, they are scattered like birthmarks and freckles over the physiognomy of our land, although most commonly in the North and West. Tough granite forms our highest peaks in Scotland and it forms the bleak wastes of Dartmoor. The influences of sedimentary and igneous rocks compete in the Welsh landscape. More subtly, the hot infusions brewed from these fiery rocks have percolated into the land itself, leaving minerals that have sparked claims and financed booms. Fire-formed rocks have been responsible for spoil heaps that are still eyesores, or old workings in remote areas that have a tinge of romance about them in craggy ruin which they probably lacked in their smoking prime.

Most geological maps are much more detailed than this, although all would recognise the fundamental three-fold classification of rocks. The basic tool then for understanding the foundations of landscape is the geological map. The names which make the map comprehensible are descriptive of rocks and of time. Colours map the rock types – or formations – as they run across country. On most maps the rocks are shown with the superficial deposits, or 'drift', stripped off. These include gravels from stream beds, wind-blown sand and the like. The maps go down to the bedrock. Occasionally the drift is the most important control on the plants that grow in your garden or the crops on a farm. Soils can vary over short distances in response to the characters of the immediate undersoil. But the broader character of the countryside is related to the rocks beneath. To a geologist a clay is a rock, for all its plasticity. Mostly, formations are named after localities where the rocks are well-developed, such as the Oxford Clay, the Bedford Sands, or the Stiperstones Quartzite. The Blue Lias and the Cornbrash are classical names for formations which do not carry a locality but have a special historical appeal. A formation can be thick or thin – hundreds of metres to less than a metre. The Cornbrash is often very thin but it was so distinctive that William Smith was able to map it as a formation on the first geological map of the British Isles (1815), and the same line can be recognised today with little change. A formation can be mapped with no idea of its age: it is primarily the rock that is mapped, not the age. A given age may include several formations in different parts of the country. On the left-hand side of geological maps there is an age calibration, which matches formation to time. But on some very small-scale maps rocks of a particular age may be coloured as if they *were* formations; this then becomes a map of time, not rock.

Geological maps are published on many different scales. The problem of representing geology is one of fractal geometry. The closer you look and the higher the scale, the more you will see: the more convoluted will a geological boundary become. It is like mapping a rocky coastline – what do you map, every protruding rock, each swirling pool? Most local areas have map sheets to a scale of 1:50,000 published by the British Geological Survey, and these give all the information the reader is likely to need. In this book we will take the widest sweep at the broadest scale – too approximate, no doubt, to satisfy

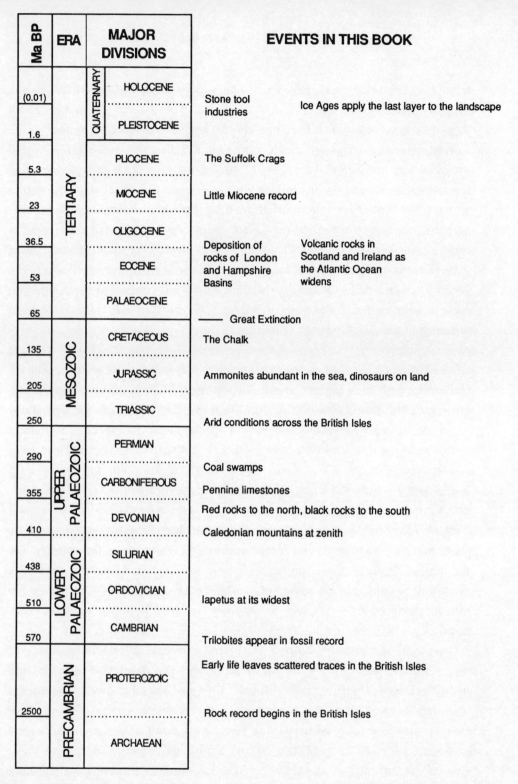

Ma BP	ERA	MAJOR DIVISIONS		EVENTS IN THIS BOOK
(0.01)	TERTIARY	QUATERNARY	HOLOCENE	Stone tool industries Ice Ages apply the last layer to the landscape
1.6			PLEISTOCENE	
5.3			PLIOCENE	The Suffolk Crags
23			MIOCENE	Little Miocene record
36.5			OLIGOCENE	Deposition of rocks of London and Hampshire Basins Volcanic rocks in Scotland and Ireland as the Atlantic Ocean widens
53			EOCENE	
65			PALAEOCENE	——— Great Extinction
135	MESOZOIC		CRETACEOUS	The Chalk
205			JURASSIC	Ammonites abundant in the sea, dinosaurs on land
250			TRIASSIC	Arid conditions across the British Isles
290	UPPER PALAEOZOIC		PERMIAN	
355			CARBONIFEROUS	Coal swamps / Pennine limestones
410			DEVONIAN	Red rocks to the north, black rocks to the south / Caledonian mountains at zenith
438	LOWER PALAEOZOIC		SILURIAN	
510			ORDOVICIAN	Iapetus at its widest
570			CAMBRIAN	Trilobites appear in fossil record
2500	PRECAMBRIAN		PROTEROZOIC	Early life leaves scattered traces in the British Isles / Rock record begins in the British Isles
			ARCHAEAN	

The divisions of geological time, providing the basis for the historical narrative in this book. Absolute ages in millions of years before the present are still subject to refinement as more is learned about radioactive decay of elements.

those with an appetite for detail – only occasionally stopping to peer closely at this interesting feature or that.

There is no end to mapping. Even now there are parts of the country where the geology is cryptic. There is scarcely an area without its mysteries, nor any area which has not at some time sparked controversy. The process of understanding the past is often compared with turning back the pages of a book to read history itself. On this analogy the rock layers are pages, the writing past life or climate, and time ticks away the narrative. But truth is much more relativistic. The writing changes, as does the tale it tells, because new interpretations are made. But the outlines of the map itself, and the stacking of time, have at last reached stability.

The colours conventionally used on the British geological map are appealing. Blues and purples are used for the ancient rocks of Cambrian, Ordovician and Silurian age, but Chalk is light green. You almost expect the rocks when you see them to conform to the colours on the map. And it is true that the older Palaeozoic rocks are often dark, though scarcely purple. The Old Red Sandstone sometimes has the brownish colour it is allotted, and the 'New Red' is suitably brighter. The igneous rocks burst from the ground in flaming red, and that seems appropriate, too. One might argue that the spring barley of the Chalklands lends the outcrop, for a while, just that jaunty shade of green. But the Carboniferous limestone, backbone of England and plateau of Ireland, is an improbable light blue. The appropriateness is obscure, unless the colour was chosen because these tough rocks formed under clear blue seas (which they likely did).

As to time, the words that are necessary to divide the Earth's history are based on the same principle as the reigns of the English kings and queens that calibrated the casing on our first fountain pen; except that there is no repetition of names in the manner of Charles or Edward, and the lengths of each division of time are less at the whim of volatile humanity. Time can be divided into segments; the divisions between segments should correspond to events in the world. Because geological time is so vast, names serve to make time comprehensible. Tens or hundreds of millions confuse: 'Jurassic' seems graspable. Even now one has difficulty in remembering just how many million years ago was the base of the Permian, but the name sticks in the memory like the roll-call of the Tudor kings and queens. The names (see endpapers) will help us shuffle events into order. It was not always so; vast miscorrelations in the past led to furious debates between scientists. Even today some creationists assert that the whole edifice of geological time is a chimera.

And it is difficult to *prove* time. Sequence is easy. Layer lies upon layer: you would need the purblindness of the last flat-earthers to deny the evidence of your eyes. As you look westwards along the Dorset coast one formation clearly dips away below another, and time follows the same design. But how much time? The convinced creationist presumably sees the evidence of successive layers piled one on another to assert God's will on the appropriate day of creation. The palaeontologist sees the time required to change one oyster to another, or to encapsulate a number of ammonite species, or as part of the history of dinosaurs and plesiosaurs: a lot of time, certainly, but thousands, hundreds of thousands, or millions of years? A scientist concerned with sediments sees the accumulation of a million tides, a thousand storms, punctuated by as many breaks of unknown duration.

The question *how much time?* has been central to geology, and still is. Time depends on the accuracy of the clock. Nineteenth-century estimates were based on totting up all the piles of sedimentary rocks – in order – and then working out how much time would be needed to accumulate them, grain by grain, by comparison with the accumulation of sediments today. Other estimates were based on the supposed rate of cooling of the Earth. The different answers were believed fervently by the advocates of their particular clock: time depended on the observer. This kind of relativity has its counterpart in the uncertain world of fundamental particle physics where the observer himself is a part of what can be defined as reality. It so happened that the most famous physicist of his time, Lord Kelvin, was widest of the mark with regard to his estimate for the age of the Earth, based on cooling, because the role of radioactive heating in the core had not been appreciated in 1897. His clock whizzed around frantically.

So what now sets the time? It is the decay of radioactive elements, tick-ticking away at a predictable rate. The click on the Geiger counter near a source of radioactivity is the very sound of time itself. Uranium (or, to be precise, one of its isotopes) throws out such clicking particles as it decays inexorably to lead (again, a particular one of its isotopes). Potassium decays to Argon at a predictable rate. As more and more precise methods of measuring smaller and smaller quantities of elements are developed, so a greater range of elements showing decay series allow for the dating of different rocks: Strontium-Rubidium, Uranium-Thorium and more. Different elements decay at various rates, so they provide clocks tuned to different timescales: fast decaying Carbon-14 dates archaeological sites; slowly decaying Uranium dates the oldest rocks on Earth. Unless there is some hidden flaw in the method which could re-set the clocks again, it looks as if the millions we now know about are nearly as real as the rocks themselves. The question of time still lies at the centre of the creationist critique. Because they *know* that the world is only about six thousand years old (the Bible and its chronology being literally true) it follows that time itself has bent, the decay of the elements distorted. Even Professor Kelvin's fast clock is too slow by far. The very atoms must be stopped in their spinning to dance to the tune of the first few words of divine revelation. Symbolic readings of the myths of Genesis are not permitted.

But the question *how much time?* is still an issue with scientists. *Rates* of change are the question: Does all evolution proceed at a slow and gradual pace, or does it go by fits and starts? Did the dinosaurs die out on a Wednesday afternoon, or slowly, over millennia? Does the Earth move today at the same pace as in the past? But the age of the Earth is now set rather securely at 4600 million years, one of those immutable facts that should be filed in the mind along with the date of The Battle of Hastings and The French Revolution. One of the important gifts that came back with the Moon rock was a confirmation of the Earth's antiquity. Earth and Moon were born together – but the Moon was not whipped through cycles of change like its giant sibling. Dead after the briefest of lives, its signature of death was preserved until men scooped its dust into their astonished fingers.

Until quite recently it was thought that the Earth and Moon were born in fire, spun from a raging sun, continuing to spin as liquid balls, which cohered and solidified as they cooled. There is an atavistic appeal to a fiery solar system, hot with youth, bubbling and incontinent with chaotic flames. It somehow seems *right* that the early Earth should be hot, and then cool to maturity, boiling off its youth in volcanoes and quakes, as purple as a fractious baby; or perhaps there is a more homely image of a hot potato, with a skin like the Earth's crust, which contains the heated interior even as it cools.

Yet it was not so. The planets grew, spun from dust and accumulated through gravity, after the 'Big Bang' of creation itself. Accretion, not fire, was the key to the birth of the Sun's companions. Even now, a girdle of such debris remains as an asteroid belt that did not congeal into planets. Then upon this agglutinated world the fire of radioactivity set in train the construction of the Earth: by heat from within. There followed the three-fold segregation – like the shells of an onion, or the flavour zones of an old-fashioned gob-stopper – into core, mantle and crust. The early atmosphere was possibly largely composed of heavy, inert carbon dioxide gas. Thousands of meteorites bombarded it. The pocked surface of the Moon preserves this phase, frozen in history; on the Moon, which soon lost any atmosphere it may have had, the craters survive. Water was expelled from volcanoes, and ultimately condensed. Metals moved in solution or in gases to find their accommodation within the crust, and if they had not, our rare and precious metals would have remained more scattered than currants in a plum duff. The Earth, in short, was mostly constructed from *inside*.

There is nothing intuitive about this tale. One might make something of the distillation of order from chaos, as anticipated in the creation myth. The heavens were indeed separated from the Earth. But here there is no comfortable image of a cooling potato, nor yet the dramatic one of a sphere of fire. Nothing here of cleaving into opposites – night and day, water and fire, good and evil – which is the stuff of myths that spring naturally from human minds. This world is alien.

But nothing remains of it. These are the truly hidden landscapes, irretrievable except in experiment or imagination, or preserved in time-frozen form on extra-terrestrial bodies. Hints of primaeval rock still arrive in the guise of meteorites, some of which are debris from that early phase of creation. The beginnings of Britain lie somewhere within this early world. Nothing of it can be seen today. The oldest rocks known on Earth are 3800 million years old. They are preserved in western Australia and in Greenland, but they contain evidence of being derived from rocks still older . . .

3800 million years . . . What can we make of such a span? During that time the surface of the Earth has been shaped, reshaped, and reshaped again, and that new shape once more re-designed. Life has appeared, diversified, suffered crises, moved from sea to land to air, developed the consciousness to learn its own past. Over the area now occupied by our small islands the sea has advanced and retreated many times. Its advances have left a legacy of sediments, from which the story in this book can be read. At other times there have been deserts, or deltas, or lakes. Is there a pattern to all this? Professor Lovelock and the enthusiasts for the *Gaia* view see the Earth almost as an organism where these endless cycles also evolve, life itself modifying the world it inhabits towards ever greater complexity and interdependence. The Earth can cope with change – even damage – because the complex balance of the planet, the legacy of its long history, acts like the immune system in our own bodies dedicated to restore balance. To others it seems as if chance alone has spelled out the history of the Earth: there is no notion of 'progress', and what we might think of as progressive change is merely a concatenation of accidents. Risen from dust, we could yet return to it.

If the truth lies somewhere between these visions this does not mean that either view is without value. It *is* undoubtedly true that the Earth as a whole has evolved since its earliest days, and that life has had a large part to play in that evolution. It is equally true that any idea of steady progress would be confounded by the intervention of chance and contingency: the invasion of the sea, the arrival of a meteorite. And especially at our local level, on our own islands, we see only a part of the story for part of the time, and that, imperfectly.

But for all that, the British Isles include as great a variety of rocks in a small area as anywhere in the world. The coming together of all these rocks – just here – is the visible expression of our several thousands of millions of years of history. The richness of our geology dictates the variety of our scenery. The stark Highlands, the crags of the Peak District, the wide tor-capped spreads of Dartmoor, the leafy Weald are all products of their geology. This is not to decry our human influence. Man has controlled the scenery in many ways, and local archaeology has shown just how persistent and pervasive these changes have been. Scholarly works by Professor Hoskins and Professor Rackham have made the reading of the landscape – Man's landscape – a popular Sunday pastime. Ancient hedges and boundaries, old coppices, forgotten villages, old green lanes and ancient industries all contribute to the details crammed into our deliciously diverse islands. But all these evidences of past activities sit on the bedrock of geology. Human endeavours do not succeed if they deny the geological realities. This hidden landscape is a part of all our lives.

3

The Oldest Rocks

To START as near as we can to the beginning, the oldest rocks in Britain are in the North-west Highlands: the Western Isles, the islands of Lewis and Harris and North and South Uist, and the coastline of dramatic inlets running from Durness to Skye. The oldest rocks come from a time where so much is obscure, and from a depth in the crust where so much has been transformed. The rocks have long been known as the Lewisian Gneiss, as if it were one rock. However modern study has revealed many rock types, and the old term should be hedged about with qualifications; this classical description is retained here only to keep the burden of new words to a minimum. 'Gneiss' is a general term applied to coarsely banded and distinctly crystalline metamorphic rocks, and the Lewisian ones can be white or pink, greenish or black depending on the particular brew of its crystals. We can only know their age from radiometric dating, and this changes as more refined methods are produced. But all agree upon the antiquity of the oldest of these rocks, which have long been known to be more than 2000 million years old. The oldest dates obtained as I write had their radioactive 'clock' set in motion more than 2900 million years ago. The still older rocks in Greenland, belonging to that Dark Age since the accretion of the planet, may then not have been so far away geographically. The metamorphic Lewisian Gneiss is the product of colossal heat and pressure. This shows in the rocks: in places they seem to twist and writhe as if to escape the grasp of a giant. In other places they are boldly banded as if modelled for coarse cloth. The dark dykes of Scourie, originally of igneous origin,

The ancient rocks of the Lewisian Gneiss, woven from the deep fabric of time . . . The most ancient, metamorphic rocks of the British Isles, Barra, southern part of Outer Hebrides.

cut through gneiss in many places, or branch and split, or are themselves twisted further. Those who study the characters of rocks, the petrologists, now know that the Lewisian Gneiss includes metamorphosed igneous rocks of many kinds. Newer dates have been obtained from some of the rocks around Laxford, which are so much younger that they show the Lewisian complex was put together over more than 1500 million years – a third of Earth's history! The Lewisian complex also includes some metamorphosed sediments, which represent a history still deeper and more arcane. Those sediments themselves must have been derived from the weathering of rocks still older . . .

The oppressions of the depth of the Earth have made Lewisian rocks hard. Yet such a span of time has passed since their forging in the depths of the crust that they do not form great mountains now. Many of the hills of Lewis, like Ben Barvas and Forsnaval, are relatively modest, as are the North Harris Hills, Visgraval Beg and Visgraval More. The Uig Hills include rounded peaks that somehow *look* extraordinarily ancient. Hundreds of millions of years of history have planed Lewisian rocks, worn and scoured and eroded them even until the roots of their creation reached sea level. Indeed, we know that this deep erosion happened even before the Cambrian period. This patch of North-west Scotland is the exhausted remnant of events so distant in geological time that much is still speculated and little is known of their world.

Lewisian scenery mostly undulates low under wide skies. Drainage is not good. There are hardly any trees now, but pine stumps are preserved beneath peat in Lewis which indicate that the climate may have favoured them in pre-history. Today the Hebridean edge of Britain is battered by gales each winter. When it is not, low sea mists often hang for hours upon the hills, and then the countryside truly *feels* primeval. Flat areas have deposits of peat which have formed from vegetation accumulated since the ice of the last Ice Age retreated, and shallow lochs like L. Roisnavat and L. Gunna on Lewis, and Lochmaddy on North Uist are acidic legacies of that time. The Lewisian scenery is governed by the conjunction of all that is most ancient, and a lot that is youngest, in the geological evolution of Britain. Its nearest equivalents are probably the 'barrens' on the Canadian Shield.

Evidence of the past action of ice still abounds, ice that probably only disappeared a few thousand years ago. Uig Lodge sits between these two worlds: the ancient migmatites of the Uig Hills on the one hand and sediments that accumulated after the last ice on the other. South of Carnish there are thick

Folds in metamorphic rocks of the Moine: the Loch Eil Psammite.

glacial deposits, sands and gravels as unconsolidated as the Lewisian rocks are hardened and annealed by heat and time. Everywhere low mounds of Lewisian rocks are polished and rounded by ice, the last erosional onslaught upon rocks that have probably seen dozens of cycles of erosion. In Glen Valtos on Lewis the road follows a channel that was cut beneath a glacier by meltwater flowing eastwards towards Loch Roag. The fjord of Loch Seaforth is a valley deepened by ice and then drowned as the sea level rose after the glaciers retreated. This recent glacial history and the uncompromising rocks below make for soil that is thin, and mostly low in nutrients. Stone walls enclose the better ground, stones lumped roughly together from those gneiss boulders left scattered by the ice, and such would also be employed to make the walls of low crofts. Tough sheep prosper, and they produce the crop of wool which is the foundation of the trade which has made Harris famous: weaving. Tweed continues to be made today, and perhaps the long winters and wild weather make it an ideal employment when outdoors is intimidating. It is not altogether fanciful to note a connection between the woven cloth and the fabric of the Harris rocks: some of the darker speckled grey gneisses can be matched by sober tweeds, and even some of the brighter rocks dotted with minerals like felspar and hornblende are a bolder weave. Curiously enough, 'fabric' is the technical term for the texture and crystallinity of igneous and metamorphic rocks, as if the heat and history of the Earth were the weaver of the rocks themselves.

And now much is known about how these most ancient of British rocks were woven. They include the record of the genesis of the Earth's crust itself. Some of the rocks were metamorphosed at depths of forty-five kilometres inside the Earth. Here, in the Outer Hebrides, we have exhumed the guts of the planet. It may seem impossible to know such details about rocks so ancient, but they preserve the fossil imprint of their own history. The 'fossils' are the kinds of minerals the rocks contain, because diagnostic minerals will only form under very special conditions. To pursue the weaving analogy, the rocks are woven from different threads, the minerals, to form the pattern of the crystalline rocks; there may be a dozen of these minerals in one rock. Very high pressures and temperatures are required to form the assemblages of minerals of much of the Lewisian 'gneiss'. These conditions can even be mimicked by experiments in geological laboratories equipped with gigantic squeezers and heaters: the minerals are 'cooked' artificially from the basic elemental ingredients, and so there is now much evidence to make this profound genesis more than mere speculation. The abundance of individual elements – rare ones like Niobium – can be measured with astounding accuracy from rocks. These elements are diagnostic: they reveal where and how rocks were formed. Hence the Lewisian history is now known to embody different phases: the deep, early phase saw the metamorphism at profound depths of igneous and sedimentary rocks. Not only this, but the whole body of rocks was sheared in the

Polyhedral garnets in a mica-schist, a typical metamorphic rock from ancient mountain belts.

great grinder of the depths, to be squeezed like toothpaste, until it gave way and fractured too. Then the entire grey, banded mass was distended, bellied upwards until it cracked, cracks which were filled from depths yet more unimaginable with magma that cooled to dolerite, a dark, igneous rock. These latter are now seen as the black dykes of Scourie, around which area on the mainland dark veins a few feet across cut through the gneisses, seemingly ignoring the banding of the older rock, as dykes do. This is not surprising, for radiometric dates now tell us that the Scourie dykes were emplaced 500 million years after the earlier events – virtually the same time that has elapsed since the Cambrian. Yet 2400 million years ago still lies far, far before the first organic remains in Britain.

Younger still, various igneous rocks were introduced. These can be seen around Laxford on the Scottish coast. Indeed, some of the metamorphic rocks were themselves heated until they began to melt, to transform into granite magma. Such 'migmatites' can be seen in the Uig Hills, where pink pods of coarsely-grained felspars and quartz seem to sit like obstinate clots interrupting the smooth weave of gneiss, or little veins of granite wander through the fabric like errant tree roots. These are the distillates of the rock itself, frozen even as they melted. True granite forms the craggy Stron Gaoith. The construction of this ancient crust continued until perhaps 1400 million years ago, when the last igneous rocks were added. This long history makes for a rich brew of rocks of many kinds. Altogether, then, the Lewisian Gneiss is the record of *more time than has elapsed since*. It is truly the ancient of days. Now dead and annealed, once it was hot and labile. In Aird Fenish in Harris the sea now troubles weaknesses in ancient gneiss to leave arches and sea stacks, and pick out faults with its probing surge. After more than 1000 million years erosion continues still, but now this crust is quiet forever. Only the cries of gulls disturb its long stillness.

Geologists refer to crust this ancient as 'basement'. This is an appropriate word, with its connotations of further stories rising above; the basement upon which all later construction rests. Later rocks, be they sedimentary or metamorphic, comprise the surface of the rest of that geologically younger Britain which lies to the south-west of Lewisian terrain, but these rocks rest, ultimately, on basement. We do not see it except in the logs of seismic shocks which probe to unseen depths. Occasional boulders of basement may be carried upwards, entrapped and plucked from their deep home and incorporated into younger igneous intrusions, showing it to be present at the foundation of everything in the British Isles. How fortunate we are to have this sliver of ancient crust at the surface in the western Highlands. Here alone we can contemplate the very beginnings of our islands.

There are parts of the world where basement rocks cover thousands of square miles: the ancient centre of Canáda, the Canadian Shield; the core of Africa; some of the flat plains of Australia. In some of these areas there are the first traces of life, preserved in sedimentary rocks, still of deeply Precambrian age, that overlie the basement but are not themselves metamorphosed. These are the record of the first inundations over newly stabilised crust. Minute fossils of bacteria and cyanobacteria now go back to more than 3000 million years. Precambrian time is often divided into two great chunks, Archaean (> 2500 my) and Proterozoic (< 2500 my to basal Cambrian). The Archaean may have been viewed once as time without life, but, as more of these tiny traces are found in places where, by some miracle, they have escaped all the recurrent vicissitudes of geological time, the division is seen as more and more arbitrary – little more than a useful label. But the traces of past life appear suddenly in our own rock successions, and it is wise to remember that elsewhere, unseen, and more slowly than can be imagined, simple plants and bacteria were preparing and shaping the world for higher forms of life: 2000 million years of preparation before the trilobite that I collected from western Wales. Preparation may not be the right word, implying as it does a plan or progression. More likely that early world was riven with arbitrariness and cataclysm, deciding which living cells survived, at some times by mere good fortune, at others by modifying themselves in the way that only living things can.

The low horizons and rounded hills of Lewisian North-west Scotland were ancient long before the first trilobite. But along the eastern edge of this country, and around the shores of Loch Torridon, there rise steep-sided mountains: Quinag, Canisp, Suilven, Cul Mor, Slioch, Stac Pollaidh (Stack Polly). These mountains lie upon the Lewisian as cakes may lie upon a table. They loom, often appearing dark, but may glow reddish in low incident light when the sun shines from the west. They can make for tough walking. But the rock is banded in quite a different way from the Lewisian: often, the banding is nearly horizontal and can be traced around the scarp with the eye, as at Beinn a'Chlaidhelmh', which contrasts with the vertical or twisted banding of the gneissose rocks. But then these mountain-making rocks are sandstones – sedimentary rocks – which have not suffered the metamorphism of the Lewisian. Rather they lie upon the eroded upper surface of the Lewisian Gneiss, indifferent to its tortured and baked history. They contain pebbles of the rocks upon which they lie, or include crystals of felspar which have been derived from the erosion of the basement itself. These sedimentary rocks are known as Torridonian.

At one time, Torridonian rocks would have provided a continuous cover over the Lewisian basement, but now Suilven and Stac Pollaidh and the rest are remnants of this great sheet, relict mountains perched on an ancient floor. Torridonian rocks spread discontinuously from Cape Wrath to Skye and Rhum. As to age, the Torridonian sediments underlie some early Cambrian rocks, and therefore must be older and Precambrian. Radiometric dates ranging from about 1000 my to 800 my prove this. A few parts of the Torridonian contain microscopic fossils of plants, single-celled algae called acritarchs, and they, too, are of the late Proterozoic, and make it likely that some of these rocks were deposited beneath the sea, although the greater part was probably the product of ancient deluges and deltas. These floods of sediment have left their traces as sand-sheets that spread from the north-west, from the direction of the Western Isles.

This means that the Torridonian sand-sheets spread over what was a Precambrian landscape floored by Lewisian, already ancient and eroded. So the Lewisian landscape that we see today is an exhumation of a far, far older landscape up to a billion years old. The rounded Lewisian hills of western Ross and Cromarty which we see now were the same hills that were lapped against by sands long before the first trilobite swam in the sea. Half veil your eyes and strip away the covering of grass in your mind, blow away the thin overburden of soil. Imagine a world with no vegetation upon the land, where violent floods swept the soil to the sea unchecked. This is the landscape that you see before you now, unveiled again by erosion.

4

The Great Divide

TAKE HOLD of a map of the British Isles and tear it in two. Start your rent in the North Sea near Berwick-on-Tweed; rip in a south-westerly direction. Tear towards the Solway Firth leaving the Lake District to the south; continue across the Irish Sea, making sure that the Isle of Man lies also to the south. Continue this line across the Republic of Ireland until you end in the sea again near the wide mouth of the River Shannon and in the Atlantic Ocean . . . You have just divided the British Isles into its two most fundamental geological components.

These two halves that you now hold in your left and right hands were once separate continents. They lay further apart 500 million years ago than Britain does now from North America. This is the fundamental division in our islands, yet it was not recognised for what it is until the late 1960s.

The narrative in this book follows geological time in an approximate way, and the basement was the appropriate level at which to start, for it has been preserved at the surface and out of time in the far north-west. But now there are suddenly two British Isles to deal with: that in your left hand and that in your right. So these halves belong, in a sense, to different narratives: like two characters in a novel that lead apparently separate lives until historical accident joins them into a single tale. But now the story must be told of how this apparently solid and immovable land was recognised to be two disparate pieces. For these islands of ours were stitched together from two great slabs of crust that fused. The oceanic crust that once lay between them vanished to leave no more than

Suilven. One of the dramatic mountains of Torridonian sedimentary rocks (Precambrian) that overlie the Lewisian metamorphic rocks in the North-west Highlands.

hints of its existence, just as a bowl that has been broken and repaired may reveal its history only by small blobs of glue along the join. Place those two torn halves of the map together again and you are re-enacting the history of the Silurian to Devonian periods; and see how difficult it is to make them join again *exactly* as if they had never been torn apart.

It is an astounding history, this fusion of the British Isles. Once, it would have been thought inconceivable. The rocks seem set immovable in their place. Mankind's follies and triumphs are the stuff of change; kings and queens are mortal, and only stone endures. In Shelley's poem 'Ozymandias' the rock-hewn inscription *Look on my works ye mighty, and despair* . . . survives while the delusions of grandeur of its subject are swallowed up in desert sands. Rock seems *permanent*: Peter was the saint named as a rock on whom the church could be founded; petrifaction, from the same linguistic root, is what turns flesh into perpetuity. That the very rocks themselves could move seemed heretical.

Yet now 'continental drift' is a familiar concept. Many people with only a glancing knowledge of science are able to picture Pangaea, a 'supercontinent' composed of all our present day continents combined as a single, vast unity. They know that the Atlantic Ocean opened, separating Europe and North America, which then moved apart over millions of years; they know that the close 'fit' of the West African coast and the eastern coast of South America reflects their conjoined past. They remember that India rifted from the other side of Africa, eventually to collide with Asia as the Indian Ocean opened up, and that Australia and Antarctica moved southwards to attain their present isolation. The concepts of plate tectonics give a scientific rationale for this history, and, even if people do not know what such tectonics are, the words are familiar. The world is composed of plates, which move, no faster than your fingernails grow. Where continents move apart new igneous crust is added at mid-ocean ridges; one testimony of this fiery construction is Iceland, where the ridges break sea level and where you can *see* how new land may grow in a matter of days in lava eruptions and flows. So our small island creeps apart from North America, carried on the back of its plate, while new volcanic crust is plastered in at the mid-Atlantic ridge. And where continents collided as a result of their movements, as India did with Asia, mountain chains were thrown up, ruckled and buckled as the crust thickened in the inexorable heave of mass against mass. But – and it is a big but – this break-up of Pangaea began perhaps 150 million years ago. What relevance has this to our torn map of Britain? For all the rocks along this severed edge are

The great divide: the Iapetus Ocean which separated northern from southern Britain 500 million years ago, probably the most fundamental geological division in our islands, and which still leaves its mark upon the landscape. This is a much simplified view of the separation; the ocean is not to scale. Subsequent closure of the ocean created the mountains of the Caledonides, and set the seal of the rocks that remain today.

far older than this. It is certainly true that to understand the Triassic world we have to close the Atlantic Ocean – to imagine ourselves marooned within Pangaea. But our story relates to rocks three times or more as old as this. And the line along which we ripped apart our islands is not the line along which the Atlantic opened. The break-up of Euramerica served to define us as we are *now*, on the fringe of Europe, and split forever from the Americas. But this may not have been how we were in the Cambrian, 500 million years ago, or more.

To explain what happened then we return to the North-west of Scotland, back to the country of the Lewisian and Torridonian.

Around the village of Durness at the coast in the far North-west of mainland Scotland there are low bluffs in the fields. Some of these little cliffs are made of tough dolomite and limestone, which come to the sea near Smoo Cave. Near the small artists' community of Balnakiel you can follow these rocks along the shore, where you can see how these rocks are bedded, and the beds form steps through which you can climb the succession of some small part of geological time. Even with the merest rudiments of geological experience you can see that these rocks are completely different from the tortured and baked Lewisian: they are hardly altered by heat at all. If you are lucky you may find a fossil – the spiral outline of a sea snail, most probably. Very rarely does a trilobite turn up, and not yet from this particular outcrop. I spent the wettest days of my geological career here. Driving rain poked horizontally to find out every crack in my clothing, producing that kind of slow dampening that creeps from knees to bum and for some reason also seems to make fossils harder to find. Dodging behind bluffs in the hope that the wind and rain would whip in another direction, I discovered my first *Drosera* – those small plants known as sundews with rosettes of leaves covered in sticky glands that entrap flies and digest them. This peaty soil contains few minerals and less nitrogen, and those nutrients carried in the bodies of flies or midges are a welcome addition to the health of the plant. Touch the leaf with a pencil and you can trigger the trap – little sticky fingers crowd in over the target. For the fly it must be like being trapped beneath a forest of falling, glutinous pylons. You can make such observations fiddling with sundews while you are passing the time thinking up reasons not to go out into the torrent again, and wondering if perhaps a career in a bank might not, after all, have its attractions. But then, a few days later, the rain stopped, and Durness was the best place in the world to be, gifted with golden beaches and that startling clarity of light that follows rain. Suddenly, to be a geologist was the only thing.

One of the trilobites that has been discovered in the Durness limestone is called *Petigurus*. It is early Ordovician in age. Only its tail is known from Scotland, and it is a distinctive object, covered in warts and lumps, and as individual in its way as a human physiognomy. The Scottish specimen was collected by the two masters of North-west geology, Ben Peach and John Horne. But I have collected exactly the same species from far away on the other side of the Atlantic Ocean, in western Newfoundland. Here, too, it is found in limestones, and the rocks compare in almost every detail with those from Scotland. But *Petigurus* is a good deal commoner there.

As I crawled over the rocks in Newfoundland the sculpted tail of *Petigurus* was not difficult to find. In fact, it was one of the largest of more than twenty species of trilobites to be found there. More than a hundred years before a gentleman called Richardson had landed a boat on these inaccessible shores and found the same trilobites. The name *Bathyurus nero* was given to one of these by Elkanah Billings at the Geological Survey of Canada in 1865, when he recognised it as a species new to science; and fifty years later the genus name was changed to *Petigurus*, because the species showed many differences from other species of *Bathyurus*. So the same species that can be found with little trouble in western Newfoundland can be found, with rather more difficulty, in Scotland. Not only that, the same species occurs in Greenland, and in the Arctic island of Spitsbergen, at about 80 degrees north on our present geography. I collected that one, too, from

The tail of the trilobite Petigurus, *one of the distinctive fossils that stitches the Ordovician of North-west Scotland to equivalent rocks in Laurentia, where it is also found in several places.*

limestones which were identical to those from Scotland. Now ice floes rub along the shores where once tropical algae and sponges lived and trilobites scuttled in the calcareous ooze.

So what are we to make of this – how can *Petigurus* help us understand the Ordovician world and its relation to Britain's hidden landscape? Nobody can doubt that these rocks with *Petigurus* were tropical, because limestones and dolomites of this kind only form under warm conditions, so they must have been near the equator at the time. *Petigurus* was as adapted to its time and place as kangaroos are to the Australian outback. So the first truth to emerge is that *this piece of North-west Scotland is really part of Ordovician North America*. It belongs with Greenland and Spitsbergen in what is termed the Laurentian plate, which includes Canada, and most of the United States at its core. The strip of this Ordovician outcrop runs from Durness to Skye, where, near Ben Suardal, other Ordovician trilobites of North American cast have been discovered, and that cuts off, as it were, the piece of Britain that doubtless belonged on the other side 500 million years ago. On the geological map it forms a thin, coloured line running from Loch Eriboll near Durness to the Sound of Sleat, in Skye. The Ordovician overlies Cambrian sedimentary rocks, which also have trilobites, *Olenellus**, which are otherwise familiar in North America, so these affiliations were no flash in the pan. Underneath the beds with *Olenellus* there is the famous 'Pipe Rock'. The pipes are vertical tubes a centimetre across produced by Cambrian worms, which must have lived together colonially in shallow sands just as lugworms do today. Now, this rock is a hard sandstone, which rather resembles one of the more exotic cheeses to be found at Sainsbury's. The Cambrian in turn overlies those rocks I described at the outset as the very foundation, the basement of Britain: the metamorphic Lewisian, and overlying Torridonian. You can see these Cambrian rocks lying upon Lewisian at Foinaven in northern Sutherland, and at Ben Arkle to the south. All this crust and sedimentary rock is, in that deeper geological reality, North America. So when we veiled our eyes and looked over the primaeval landscape of undulating Lewisian we were seeing some sliver from the edge of the great Precambrian Canadian shield, stranded by time's accident. That is the hidden truth concealed in this north-western corner of the British Isles.

Is it a coincidence that Scots have colonised Canada so successfully? Or perhaps the starkness of Precambrian Canada was less intimidating to these people in particular; perhaps the Celtic settler was in a sense returning to a deeper reality.

But now we must go some way to restore things to how they were – by closing the Atlantic Ocean. We lose all that oceanic crust which grew at the Mid-Atlantic ridge, turning time backwards to suck the two halves of the North Atlantic together again. This is to step back to the Permian world. And now, of course, our Scottish limestones with *Petigurus* are much closer to western Newfoundland – and to Greenland and Spitsbergen. On present geography this is nowhere near the equator. But then this ancient world with its moving and redistributed continents can be reorientated with respect to lines of latitude, so perhaps that presents little problem.

What I have just related is how the story of Britain's structure might have been presented to the reader in the mid 1960s. But it has missed the most important fact of all . . . the rent across our islands that began this chapter.

Let us return to *Petigurus* and the tropical world to which it was adapted.

---

*It would be wrong to give the impression that this awareness of the North American affinities of rocks and fossils is new. The pioneer geologists of the last century drew similar conclusions from studying rather less distinctive fossils than *Petigurus*.

*The Pipe Rock, near Durness, Sutherland, a Cambrian sandstone much burrowed by ancient 'worms',
which have left their traces crowded together as vertical tubes.*

When we closed the Atlantic we joined *Petigurus* once more to its brothers and
sisters and friends in Laurentia, and in so doing restored this piece of Scotland to
its proper place. But in the process we also dragged behind it the *rest* of Britain,
to say nothing of Europe, lying to the South. Back in the Ordovician, should this
area not also show sure signs of relationship to Laurentia if it lay so close?

The geological map shows plenty of Ordovician further South of the Laurentian
strip. If we take a geological hammer to these rocks, and if there are trilobites
to be found, we should perhaps expect to find *Petigurus* or one of its associates.
The Lake District may be a good area to try because there are plenty of tumbled
rocks on scree slopes, or outcrop exposed in stream beds, much of it exactly
contemporaneous with the carbonate rocks of Durness. Look at the Dead Crags
or Randel Crag around Skiddaw, or at Outerside. But at once you can see that
the rocks are different. No limestones these, they look like the slates on your roof,
at least they do if you happen to live in a Victorian house. They slide in slithery
heaps down Randel Crag. If you can struggle up the scree slope you will find
the outcrops: no stepped beds like those on the beach at Durness but crammed
and interleaved like petrified playing-cards. Like cards, their surfaces are flat. At
Durness, you cannot break out flat slabs from the tough dolomites, but here it

is hard to break the rock any other way: it demands to fall into sheets. Clearly, something has happened between Durness and the Lake District.

To find out what, we must search for trilobites again. If you are very lucky, you may find one, and if you do, it will lie on the surface of one of the sheets of slate like a kipper on a platter. But over the last hundred and fifty years no *Petigurus* has turned up, though the Ordovician stretches for dozens of miles. Nor has *any* Laurentian trilobite been found in these slabby rocks, although twenty different kinds have been found. Names, those names that label time and place, are all different here. To use a crude analogy, if you found yourself in a strange town in a strange country, and had only the use of telephone directories to find out where you were, and if you found nothing but Spanish names, you might conclude that you were not in North America. The names of trilobites from the Lake District are familiar elsewhere, but not from western Newfoundland, or Greenland, or the USA, like *Petigurus* and its friends. Rather they occur, as in our analogy, in Spain, but also in Wales, and Shropshire, in France, in Morocco, in Bohemia and Portugal, even in Libya, from dozens of localities. Clearly the Lake District was part of a huge area of Europe and North Africa, or, to keep it close to home, like Wales and Shropshire, but not like North-west Scotland.

I have spent many hours in Welsh ditches looking over the surfaces of shales for trilobites, and I have never discovered one species known from North America or Scotland. Dozens of earlier investigators have done the same, but among their collections there are no North American species. In total, thousands of hours must have elapsed in the splitting of tons of rock in the search for early Ordovician trilobites – but in Wales there are no strangers from Scotland, nor in Scotland any strangers from Wales. Nor are there any limestones or dolomites of this age from Wales, indeed they are absent over the whole of contemporary Ordovician Europe and Africa. Since we know that these limy rocks were formed in the tropics, does their absence not suggest that the trilobites may have lived away from the tropics in this Lake District-Welsh-Euro-African area? It seems that *Petigurus* may be telling us something important.

The answer seems, in retrospect, obvious, as so many brilliant answers do. The Canadian geologist J.T. Wilson published the essence of it in a short note in the scientific journal *Nature* in 1966, entitled, simply: 'Did the Atlantic close and then re-open?'

Maybe, went the argument, the differences between Wales and the Lake District, on the one hand, and North-west Scotland, on the other, were because they lay on either side of an Ordovician ocean. *Petigurus* simply could not reach these Anglo-Welsh trilobites, nor they *Petigurus*. They were two different nations, and their names, and the distribution of their names, reflected it. Then this Ordovician ocean closed, just as the ocean which once lay between India and Asia closed. It closed when the supercontinent Pangaea came together. For Pangaea itself was only a phase in the history of the Earth, assembled from drifting continents from a still earlier stage. Back in the Ordovician we were in this early phase, where we have to learn geography all over again. The Earth is in continual motion, it was implied, with continents drifting apart or coming together as the oceanic crust between them is created or consumed.

So where did this ocean run? It could scarcely have involved only the area between the Lake District and Scotland. Once again, if you close the Atlantic to its Permian-Triassic position you can see at once that this area of Scotland continues to the North as the Scandinavian mountains, forming much of Norway, and to the South as the Appalachians, following the obvious north-east–south-west trend of the rocks. This was a vast mountain chain, now much worn down by hundreds of millions of years of erosion. Suddenly, it is as plain as a pikestaff: the whole Appalachian – Scottish – Norwegian chain is an ancient Himalaya, or a fossil Alps. Like these mountain chains, it must have been produced by the collision of continents. But this collision happened when Pangaea itself was assembled by the coming-together of continents, for the chain runs within the supercontinent in the same fashion as the Himalayan chain runs through Asia. Thus, the Ordovician difference between Durness, with *Petigurus* and being a part of Laurentia, and southern Britain, without *Petigurus* and part of a European-African continent, is on account of the interpolation of a vanished ocean. This ocean has been called Iapetus. Later, this ocean disappeared by subduction, and when the continents approached one another, a mountain chain, huge as the Alps, was thrown up. This chain, the Caledonides, ran the length of the North Atlantic as it is today. Continental collision brought closely together fossil animals that had once lived separate lives in separate seas. Welded together in Pangaea, the two continents were spliced for 200 million years, while erosion worked upon the Caledonides and pared them to their roots. Then Pangaea itself began to split up 150 million years ago to give us the arrangement of continents we know today. And so the Atlantic Ocean opened – or rather it re-opened along the line of weakness generated long before by the closure of Iapetus. Where else? But it did not follow the old lines *exactly*. Instead, fragments of Laurentia were stranded on 'our' side, and elsewhere bits of Europe were stranded on 'their' side. That is why North-west Scotland is Laurentia in that deeper geological reality, and why we find *Petigurus only* there.

So we move from a trilobite to a continent to an ocean. When J.T. Wilson published his suggestion he used particularly the evidence of the animals. But it soon became clear that the vanished Ordovician Iapetus Ocean and its consequent mountain range explained so much that had been obscure, and placed in a logical frame so much that had been arbitrary. Here was the essence of our hidden landscape. Why, it was obvious that the simplest fact about the British Isles was the great swathe of metamorphic rocks of the Highlands, and this is where the roots of the Caledonides were exhumed. The granites of the Cairngorms welled up in those very roots. So much falls into place.

And the rent with which we started? This is the line at the northern edge of the European continent, a line which many geologists also believe marks the site of the disappearance of the great ocean of Iapetus. When I visited the Solway Firth I confess that I felt disappointment. If this is where an ocean disappeared, there is not much to see for it: a stretch of sea water, some flat ground and rather dull countryside before the swell of the Southern Uplands. Geophysicists have found at depth the great northward-dipping plane where once Iapetus departed by subduction beneath the area we know as southernmost Scotland. And perhaps this is enough. But somehow one expects to see something on the ground comparably dramatic to the story I have related: a great rent in the Earth, perhaps, where even now earthquakes might remind us of the past that in the Devonian had mountains towering thousands of feet above where we stand today, and where the great cycle of the world and its crust took one of its crucial turns.

This then is the great divide. Traced across the Irish Sea it follows the Caledonian trend (NW–SE), but its course there lies partly concealed beneath the younger blanket of Carboniferous Limestone. If 'continental drift' had seemed not to relate to us in our small island, being something that went on in Africa, or Australia, on the contrary, here is its deepest stamp on the fabric of our land. Ancient continents have drifted, come together, and parted again, and this leaves a legacy that even now affects the soil and vegetation and livelihood of the British Isles. It is a marvel that this division has only been recognised for what it was so recently, for all that Scottish rocks have been discussed in a thousand ways from the earliest days of British geology.

That the continents continually move and reconstruct themselves in new patterns seems curiously appropriate to our relativistic times. The fixity of the Earth has been lost. What the surface of the world looked like depends on where you are in history. Even Pangaea was merely a phase. This reflection should be made as you stand on a high crag in the Lake District and look northwards to where an ocean vanished. Furthermore, because the reconstructions are based on the inferences of scientists, fallible and imperfect men and women, the form of the prehistoric world is intimately tied in with the personality of the observer. One scientist 'sees' the ancient world in a different way from another – and who is to say which one is right? This is like the uncertain world of fundamental particles at the ultimate and smallest scale of matter. Particles are waves, or discrete entities, or measures of probability, or all of these depending on circumstances and whoever is making experiments upon them. It is a will-o'-the-wisp kind of universe where everything is simultaneously, and equally truthfully, also something else. The fixity of the rocks may have seemed a sheet anchor in this storm of uncertainty, but now even geology is part of the maelstrom. The rocks upon which Ozymandias inscribed his name are themselves the stuff of change. The crust cracks, and buckles, and the continents slide around the world; new oceans open up between them, while old ones die as their oceanic crust plunges downwards at subduction zones. The world is ever in transit.

It may have been wrong to start with the Iapetus story. A clever, histori-cal account might lead the reader through the host of past interpretations of Precambrian to Devonian British geology, before pulling Iapetus like a rabbit out of the hat to show how its former existence explains all. By revealing the denouement now I have little choice but to stuff the rabbit back into the hat again. But it *is* true that the concept of Iapetus has made us look again with new eyes at British rocks older than Carboniferous in age. Nothing now looks quite the same. For the way we understand landscape has to do as much with what is in our minds, as with what we think we see with our eyes.

In North-west Scotland, on the road south from Assynt to Ullapool, near Elphin, you can follow a marked nature trail to Knockan Cliff. Here you can see, in several places, a clear succession of rocks. Recalling that in simple sequences older rocks lie beneath younger ones, there seems no problem with the geology here. At one point you will find what look like horizontal beds overlying rocks which are reminiscent of the Ordovician Durness Limestone, a name which will be familiar to you. Thus you might conclude: here are some rocks of Ordovician age, or perhaps younger, lying over the Durness rocks. Look a little closer at the

Durness Limestone and you will notice that it has a kind of unhealthy look – the bedding less clear than at Smoo or Balnakiel, and sometimes smeared, but then they are very ancient rocks and so perhaps we should not be surprised at signs of suffering. Now look more closely at the beds overlying the Durness rocks. There is something odd about them. With historical hindsight we might recognise the typical features of metamorphic rocks – the shiny surfaces covered with little mica crystals, for example – though this may not have been apparent in detail to an early observer. These are the rocks of the Moine.

This contact, between the Moine 'schists' and those rocks underlying which were known to yield fossils, was the subject of one of the great intellectual struggles of the last century, at a time when the structure of Britain was being elucidated for the first time. For these Moine rocks lie at the western edge of a great tract of difficult rocks for which fossil evidence of age is lacking: solve the age of the Moines and you shine a great beam of light into the past of the whole of Scotland. Small wonder that the Highlands Controversy was fiercely fought! David Oldroyd (in *The Highlands Controversy*, 1990) has meticulously reconstructed the battles from the contemporary documents, thus avoiding the 'Aha! if only they had known . . .' kind of history. Because honest men made honest errors on this complicated ground, just as wilful ones made precipitate judgements, the person-alities of those involved played as important a part as the exigencies of society. Sir Roderick Murchison was doubtless a great geologist, but he was a domineering aristocrat to boot, who was not above using his influence to further his ideas. He conceived in the 1850s that the succession I have just described at Knockan was a normal sedimentary one. So he thought that the Moine Schists followed the fossil bearing sequence to the west, just as Cretaceous follows Jurassic in the South-east of England. This permitted the whole of the Central Highlands to be recruited into Murchison's own 'Silurian System'.

Murchison coined the 'Silurian System' to embrace what we would nowadays include in Cambrian, Ordovician *and* Silurian Systems. James Nicol, originally Murchison's travelling companion and fellow investigator, and later professor at Aberdeen, dared dissent, seeing evidence of violence and movement all along the line from Durness to Skye. Poor Nicol! He was no match for Murchison in cutting debate, nor in the promulgation of his ideas, the more so when Murchison recruited Archibald Geikie as his ally.

In secondhand bookshops in Rye, or Croydon, or Peebles, there is usually a small section devoted to scientific books. *Gray's Anatomy* is always there, shoving aside less momentous works through sheer skeletal and visceral bulk. Tatty and unloved, old editions of *Gray* have been replaced for popular use by multicoloured compendia of symptoms produced by the *Reader's Digest*, or, one supposes, by the latest edition of *Gray* (for the serious student). An even smaller half shelf may have geology books, and one of the books that is always there is Geikie's *Text Book of Geology* ('2 vols, Octavo, Gd. Condition, slight foxing'). My own copy is bound in a dingy shade of green. But open the covers and read. Here, you soon discover, is a real writer, with a gift for explanation that makes rocks comprehensible and geology dramatic. No niminy piminy academic dryness nor technical obscurantism fuddles the description of the world. It is no wonder that the book ran to many editions and reprintings, nor that ordinary citizens of Croydon, or Rye, or Peebles bought a copy to learn about geology.

Geikie was probably the most successful career geologist of the nineteenth century, for he had not only literary talent, but also scientific acumen, astounding energy, and, not least, political nous. The latter ensured him the chair in geology at Edinburgh, and the Directorship of the Geological Survey of Great Britain and Ireland, the post from which Murchison had wielded so much influence on geological thinking. Geikie's advancement was not unconnected with his close relationship with Sir Roderick Murchison. The chair in Edinburgh was effectively created for him with Murchison's help. In his early career he explored the North-west Highlands in company with Sir Roderick, and evidently agreed with, and promulgated with typical eloquence, the Murchisonian explanation of the Highlands. Nicol's several attempts to associate the contact of the Moine Schists with what lay beneath with earth movements came to nothing. In public confrontation and in professional publication he was dismissed by Murchison as troublesome, but defeatable. He lost. 'Is it then wonderful that along this line the strata should be crushed, contorted, thrown into apparently discordant positions?' he wrote in 1866, before finally leaving the controversy alone, a defeated and disappointed man.

Well, yes, it was indeed wonderful. And many other geologists came to examine the Moinian Line near Durness, or Assynt, or Loch Eriboll, where the rocks were exposed that allowed them to form their own judgements. Some of these visitors formed opinions which were not in accord with those of Sir Roderick Murchison. The sheer complexity of this line was undeniable, for sections seemed to repeat, and in places the Pipe Rock, with its distinctive worm tubes, was doubtless turned completely upside down . . . One can visualise the small inn at Kinlochewe, at the head of Loch Maree, buzzing with speculation and discord.

After Murchison's death, in 1871, the dissent was aired more in public. The 'professionals' of The Geological Survey came to be ranged against 'amateurs' – which in those days embraced university men or museum curators alike. The Survey continued to defend the comparatively simple succession outlined by Murchison, but by the 1880s the barbs were beginning to lodge in the target. Geikie had taken over the directorship in 1882, and was slow to yield to criticism.

It was two palaeontologists, Charles Callaway and Charles Lapworth, who provided the basis for modern interpretations. I collected one of the small tri-lobites named by Callaway, *Conophrys salopiensis*, for scientific study as this book was in gestation. He had discovered this trilobite in Shropshire, which was really his home territory, and it is a measure of the lure of the Highlands controversy that he took himself off to Scotland in 1881–3, then no easy endeavour, to see things for himself. Callaway examined the sections in greater detail than had been done before, and proved the overturning of the rocks in places along the line of contact. He drew in faults with a low angle of inclination, juxtaposing unlike rocks on top of one another. He conceived the essence of the solution. Lapworth was a masterly geological mapper, and as he mapped the area around Loch Eriboll, he saw how the contact beneath the Moine Schists rested here on one of the underlying sedimentary formations, there on another. The rocks at the contact were ground and altered. He made the intuitive leap that marks dis-covery: the Moines had moved, pushed from the South-east, *over* the limestones and dolomites and sandstones, even the older Lewisian Gneiss. He wrote to his colleague Bonney (in 1882):

> Conceive a vast rolling & crushing mill of irresistible power, but of locally varying intensity, acting not parallel with the bedding but obliquely thereto; & you can follow the several stages in imagination for yourself. Undulation, corrugation, foliation and schistose structure – slaty cleavage are all the effects of one and the same cause . . . Shale, limestone, quartzite, granite and the most intractable gneisses crumple up like putty in the terrible grip of this earth-engine – and all are finally flattened out into thin sheets of uniform lamination and texture.

Thus Lapworth dramatically encapsulated what happened at the edge of our Laurentian fragment. The simple sedimentary succession was banished forever.

As in parts of the Alps, the metamorphic rocks had been forcefully *thrust* over the ancient crust and sediments to the west, but at a low angle, thus to mimic a sedimentary succession, a duplicity of nature that had persuaded Sir Roderick. Like many satisfying explanations it seemed to account for much of what had previously been inexplicable at a single stroke; old and puzzling observations reshuffled suddenly formed a coherent pattern: now we *knew* why metamorphic Moine Schists sit on top of unmetamorphosed Cambrian sandstones, and why the rocks beneath the thrust seemed in places as if ground in a mill. The excitement of confronting Geikie may have been too much for Lapworth to bear; in any event his health broke down in 1883 'feeling the great Moine Nappe grating over his body as he lay tossing on his bed at night' as Sir Edward Bailey described it. But by 1884 Geikie had publicly recanted, faced one might say with evidence beyond refutation. It could not have been easy for one so powerful and persuasive.

But the 'professionals' redeemed themselves. Geikie, by then Sir Archibald, as if as a recompense, put two of his most brilliant field geologists on to the North-west Highland problem: Ben Peach and John Horne. Together, they produced one of the masterpieces of geological mapping, with an accompanying memoir published in 1907, but summarising much that had gone before: *The Geological Structure of the North-west Highlands of Scotland*. This is still consulted by those who traverse the area. With it, and after fifty years, Murchison's theories at last became the province of the historian. The great surface over which the Moine Schists slid or scraped across the Lewisian and the sediments that lay upon it was The Moine Thrust, but it was shown to be a family, if not a tribe of thrusts, responsible for the complexities that had baffled earlier observers. In the end it is fitting that two survey officers should have at Inchnadamph, where their ghosts may look across at Quinaig, a plain stone memorial inscribed: 'To Ben N. Peach and John Horne who played the major part in unravelling the geological structure of the North-west Highlands 1883–1897.'

Now back to Knockan Cliff: we might see the same rocks, but how differently! Is it not, after all . . . well, obvious? Even the casual visitor may saunter up to the rock face and airily wave his hand. 'The Moine Thrust', he will observe, insouciantly. Root around a little and it is evident that the Moine Schists come to rest, not just on Ordovician Durness carbonate rocks, but also on the underlying Cambrian sandstones, and other rocks of that fragment of ancient North America, depending on how far westward the thrust plane extended. In places where the rock beneath was ground to a paste by the massive north-westward

A boulder of folded mylonite — a 'crush rock' ground beneath the Moine Thrust, Ross and Cromarty.

heave of the schists above, the legacy of movement is preserved in banded rock which Lapworth called mylonite. Such rocks are best seen, not at Knockan Cliff, but at the eastern side of Ben Arnaboll between Loch Eriboll and Loch Hope, where you can find blocks lying among the grass, stripy as regency wallpaper.

It is as well to recall that this story is only retrospectively obvious. Its unravelling caused decades of angst, and spoiled years of the life of an eminently honest Aberdonian professor. Nowadays you can easily drive along this stretch of North-west Scotland, for the road almost follows the Thrust. Listen hard: you can almost hear the grinding of rock mass against rock mass; and, perhaps also, faintly, the crash of will against will of those who unravelled the secret of the landscape. Even now, if one is honest, is it not difficult to conceive that one can place one's hand upon the very plane where heaving mountains ground against the immovable mass of the Laurentian shield? One somehow expects the drama, on the ground, to match that of its history and the unravelling of that history, but the track of the Moine Thrust is rarely that momentous: a change in slope, a dip between bluffs. One almost sympathises with Sir Roderick: can this *truly* be the line of one of the greatest convulsions of our islands?

Once the Moine Thrust had been christened, once Lapworth had named the crush-rock as mylonite, it fixed the truth. This was no mere exercise in nomenclature, this was the prescription of reality. No jargon either, for the names of this reality were purchased with the tramp of boots through bog, and the slow delineation of a thousand sketches. Men's reputations hinged on it, and it lay behind the turmoil of how many sleepless nights and brooding days.

Wordsworth evidently disdained the geologist who named names:

> He who with pocket hammer smites the edge
> Of luckless rock or prominent stone,
> . . . detaching by the stroke
> A chip or splinter to resolve his doubts:
> And with that ready answer satisfied,
> The substance classes by some barbarous name,
> And hurries on . . .

I suppose what Wordsworth had in mind was the kind of dull naming whereby to give a name is all that matters – a kind of rock 'twitching'. Such nomenclatural business may have seemed to him only to eliminate poetry, or, rather, to gloss over the reality that was accessible to the poetic sensibility alone. But there is something about the Moine Thrust that *is* almost poetry. What could be more dramatic than the grind of rock against rock beneath the terrible grip of a vanished mountain range? In this case a name is shorthand for a vision. And who could doubt that to see upon the ground the vestiges of a distant past adds to the richness of our experience of the present? There is an exquisite irony that sheep, the most nervous of animals, now peacefully graze slopes where continents came to rest. We may see only the ooze of a small, rush-rimmed spring to mark where rocks of unlike type came to lie one upon the other. The subtle differences in permeability of the rocks recognise the truth, where the ignorant walker could pass by unenlightened, and the wind blows in the cotton grass as if none of this had ever been.

Thus was the Moine Thrust line seen to mark the edge of the great Caledonian mountains to the East, now worn down to their roots. Had you been standing on top of Ben Hope in Sutherland forty years ago with a book like this to guide you, you could have looked south-eastwards and reconstructed in your mind's eye the vanished peaks of this lost Alps, and westward you would have known that the coastline there was Lewisian crust that was already ancient when these

Alps were young. Somewhere between, a thrust line – dimly seen – marked the impinging of these two worlds.

What you would *not* have seen is the reason for this alpine chain. Because at that time it was not realised that Iapetus had been there. The vanished ocean with which this chapter started is the missing piece of the puzzle.

Iapetus explains what needs to be explained: why there was a mountain chain to the east of the Moine Thrust. Just as we have stood where the Lake District is now, looking northwards into a vanished ocean, so we could have stood in the Cambrian or Ordovician somewhere along where the Moine Thrust runs now, towards the edge of the Laurentian continent, and seen an ocean stretching away from the tropics and beyond the horizon. The southern half of Britain would have been a distant dream, as one may stand at Land's End and dream of Manhattan. In the Ordovician, we would have stood on a limy flat, and *Petigurus* might have tickled our feet. If we had been hungry we might have been able to prise open the lid of a fat but sluggish snail called *Ceratopaea*. It would have been hot, and we would have been grateful to paddle in small pools between cauliflower-like 'heads' made by blue-green 'algae'. Our feet would have stuck clammily in thick lime mud (destined to be limestone or dolomite). Just possibly, in the distance, there may have been visible a few low islands of Lewisian granulite.

But the ocean closed. And as it closed the sedimentary rocks that lay off Laurentia, together with archipelagos, and islands, and tracts of all-but-vanished ocean crust, and much mystery besides, piled up and buckled, and slid and thickened into a mountain chain, our Scottish segment but a morsel of the whole that ran from Alabama to Tromso. The motor that drove Lapworth's 'great earth machine' was the same motor that powered, and powers still, the rise of the Himalaya and the terror and grandeur of the Alps: the movement of the plates as oceans disappear through subduction and continents collide.

And the edge of all this was the Moine Thrust. It was the final heave westwards on to the rigid mass of Laurentia. With what we know now, it is not surprising to learn that the Moine Thrust comprises a whole complex of thrusts: the edge of such a momentous crustal confusion is unlikely to have been clean. It is now also known, from the evidence of radioactive minerals, that the thrust zone had a long history of movement besides – there was not just one mighty thrust. This is the stuff of detailed geological snuffling using the whole battery of modern techniques. But the important thing is how our perception was shifted by the marriage of Lapworth's thrust with the vanished ocean, Iapetus.

The North-west of Scotland is cromlech country. Perched on hills, they comprise three upright stones and, resting on top as the seat does upon a stool, a large, horizontal stone slab. They were once neolithic burial chambers, long since robbed of their covering of earth. Cromlechs seem almost as remote from us as do the Caledonian events I have been narrating, even though they were made by the hands of early human settlers. They are perhaps one three hundred thousandth as old as the Moine Thrust, if one is silly enough to put a notional figure on it. Yet there is a curious metaphorical similarity between these prehistoric manmade structures and the discordant rocks, which lie in contact with, yet at odds with, what lies beneath. Our minds move to new thoughts by lateral leaps rather than deduction. Is it possible that the pioneer geologists contemplating the chaotic Moines translated an image of a cromlech upon the land itself?

On the north coast near Durness, there is a wonderful cave within the Durness limestones called Smoo Cave. I cannot leave North-west Scotland without quoting a description of this cave by Matthew Heddle (1882):

> The light which slumbers within that dome, once seen, is *felt*. The light – what they have of it – of all caves is fine; that of this is surpassingly lovely. Poets would call it a *chastened* light. If by that, is meant that it resembles a character which, through the buffetings and disappointments of the world, shines with a softened sweetness, we partly understand the application; and it would be a fitting one.

This from a professor of chemistry at St Andrews University! I wonder if Wordsworth would have approved? Certainly, he could not have accused Professor Heddle of being a mere dubber of dry names.

5

'Here be dragons': Caledonia

THE TOWN of Callander sits on the northern side of the Midland Valley of Scotland, snuggling down on the edge of the slopes to the north as comfortably as an elderly aunt settling in front of the fire. Reassuringly solid and sensible stone houses are designed to face out bad weather when it comes. Hotels dot the hillside and command a sunny southward view. Along the main street Scottish wool shops offer tartans. Tea shops are interspersed. Scones and jam are briskly served on clean table cloths to sensibly dressed women in tweed skirts who come out on day trips from the suburbs of Glasgow. To the south there are fields, and cows, and broad-boled oaks at the edges of fields, and a comfortable look to the countryside.

How different it is to the north! The road through the Pass of Leny climbs dramatically. Lorries grind upwards in their lowest gear. At first there are a few houses perched precariously, and then there is little overt sign of suburban humankind. Birch woodland, or small oaks with twisted, mossy branches line the steep valley, and the river that has carved it bounds over its bed. Clearly, we are in a different land: the geology has changed, and with it every aspect of the landscape. The road north has taken us across another of the great defining lines of British geology: The Highland Boundary Fault. On the geological map it is clear and straight as a crack in a window: to the south the Midland Valley, to the north the Highlands. It has the North-east–South-west trend that we would now think of as Caledonian. Just as the ground climbs in height so it descends

The Great Glen, the site of a huge fault which may have been the match of the San Andreas Fault in California. View NE from Banabie, showing Loch Lochy, Loch Oich and Loch Ness. The Glen was scoured further by glacial ice during the Pleistocene.

in time. Much of the land to the north belongs to the Precambrian. It is the highland of the high Caledonides. Sandwiched between the Moine Thrust and the Highland Boundary Fault is a great slab of land that has something of common character, not least a certain toughness. This is where population density plummets, and where the Gaelic language lingers, in patches. This is the country where metamorphism rules.

The Highland Boundary Fault is a deep cut. In contrast to the wandering thrust of the Moine it slices downwards, truncating the geology sharply. Because the younger rocks are to the south, it follows that this southern side is the one which has been let down – it is the 'downthrow' side. But the topography along the fault is not only because of the throw; it is defined by the difference in resistance of the rocks to weathering on either side of the fault, expressed over millennia. The more resistant rocks lie to the north of the fault where they take the high ground. Such is the country of bens and lochs, where millions of years of erosion have been compounded by ice scour 'merely' in the last million.

One loch crosses the Highland Boundary Fault at its western end: Loch Lomond. Although its place in Scottish mythology is assured, it is truly anomalous. It is wide and irregular and dotted with islands at its southern end, narrow and straight to the north. It partakes of the two geological worlds to either side of the fault; the northern part is a typical Scottish loch lodged in the Highlands, narrow and deeply shelving and ice-scoured, while the southern part is an altogether gentler *lake*, where you can paddle in places, and yacht clubs have a chance of making a profit.

Near Balmaha, on the south-eastern edge of Loch Lomond, you can walk across the Highland Boundary Fault without working up a sweat. The road northwards around the loch takes you successively through the rock formations: but the fault itself is scarcely more than a dip in the road. Just south of the fault it is easy to spot a rock composed of pebbles and boulders plastered together in a reddish matrix, looking like a particularly coarse concrete. This is a *conglomerate*. That it resembles concrete is no coincidence, because it too was formed from a collection of cobbles of various sizes cemented together. This conglomerate is tangible proof of the former existence of the Caledonian range lying to the north, for it is composed of the fossilised debris that tumbled, or was washed by brief torrents, off those vanished peaks. Stacked like fans against the fault scarp this is sedimentary rock at its most unsorted, a pot pourri of irregular bits of rock and slabs and cobbles all crudely dumped together during the Devonian.

Conglomerates produced from the wasting of the Caledonian Mountains, the coarse debris from a deeply eroded mountain chain. Lower Old Red Sandstone SW of Meg's Craig, Auchmithie, Tayside.

The fault then was active: the mountains rose and were stripped, unsoftened by greenery, naked to the fiercest of elements – the rate of erosion must have been staggering.

North-east from Balmaha a 'waymarked path' leads to Conic Hill, the profile of which requires no description. As you recover your breath on the way there are plenty of opportunities to examine some of these chips from a vanished mountain range – break them, and they may well be glistening metamorphic rocks which betray their pedigree. Some pebbles you can still pluck out from their matrix as you might an almond from a cake. They can let you down if you trust them for a foothold. But in spite of such betrayals it is not difficult to scramble to the top of Conic Hill, and there, on the boundary between two geological worlds, look north and west. For a moment you could be forgiven for thinking that the great Caledonian range was still there, as peak after peak retreats into the distance, those further away obfuscated by mist to suggest unseen regions still more precipitous. The valley just beneath you tracks the Highland Boundary Fault, which erodes, just as all cracks erode. Between this and the Moine Thrust, which can be imagined but not seen, is the great wedge of the Highlands, the most complicated area of geology in Britain. In your mind's eye follow the Caledonian (NE–SW) trend across the Irish Sea and this slab of geology extends further – into Connemara, North-west Ireland.

But this alpine aspect is a chimera. For although the ruggedness is real enough, most of the land before you is under 1000 metres. Ben Nevis famously tops 1344m; most of the other peaks which approach this height are in the Cairngorms. This is a fraction of the height of a young and vigorous alpine mountain chain – one-sixth of Everest, one quarter of Mont Blanc. These great corrugations of the Earth start where the Highlands stop. Three hundred million years of erosion have worked and worked at the Scottish peaks, seeking out every weakness, exploiting every crack. Rain has dissolved what it could. What it could not has been the victim of buffetings from a hundred million gales. Water has insinuated itself into the finest hairline cracks within mineral grains; when it has frozen it has prised the very minerals themselves apart. A wedging, chipping, freezing, blowing, dissolving assault has been perpetrated upon the relic of Iapetus until its high peaks have vanished into the endless mill of the Earth. Most recently, the whole has been buried beneath a great blanket of ice, which has carved and gouged and fluted the surface of the landscape.

More has been removed than remains. What we see now was once at great depth beneath these vanished Alps. From the vantage point on Conic Hill you would have to cast your eyes upward into the sky until you crick your neck to truly envisage the Palaeozoic peaks, and to estimate what has been removed by erosion. Yet what remains still stands higher than any land in Britain, because such alpine rocks endure. They have been proved in heat and annealed under pressure. There are granites of implacable toughness in the Highlands. The grinding of glaciers does wear down granites, but slowly. There have been periods in the past when the rocks have been ground almost to a plain, only to be uplifted further so that the cycle of erosion can begin again. It may take another 500 million years to pare down these ancient mountains to the deep level that has been reached by the Lewisian.

But with so much removed by erosion, it is no wonder that the geology of the Highlands has been so difficult to read. It is a story with chapters ripped out, and written in a strange tongue. Great brains have pondered it. Yet, until recently, vast stretches of land remained obscure, and beyond understanding. Some still are. On my first geological map the northern Highlands are coloured yellow, and carried the legend: 'NOT SURVEYED IN DETAIL'. This is a sober equivalent of 'HERE

BE DRAGONS': a mysterious land of Celts and crystals where few roads cross. This is the countryside of Cnocs and Meills labelling a hundred pointy hills. To a palaeontologist like myself it is a strange land without fossils, where the signs we use to read the past are suddenly unavailable. It is easy to get lost. There are, to be sure, a handful of trilobites that escaped metamorphic mangling, which are now found just to the north of the Highland Boundary Fault. These are friends of *Petigurus*, which show that the metamorphic Highlands must have lain off the edge of Laurentia in the Ordovician. But metamorphism has overprinted the record of the past over the vast waste of crystalline mountains, and removed much of the evidence of fossils that there might once have been. The detective of the past must seek other signs, hidden in the minerals, the fabric, and the structure of the rocks.

The great granitic masses of the Cairngorms, Braemar, and Donegal in Ireland are the most uncompromising of the ancient rocks of the Caledonides. Granite* endures as the most obdurate of igneous rocks. These great *intrusions* welled up as magma into the roots of the great mountain chain, mostly at a late, Silurian stage in its evolution. Their exhumation is proof of the depth to which erosion has carved. The crystals of granites interlock like a three dimensional jigsaw puzzle. The fretsaw of Pluto can dovetail pieces together with a complexity Man cannot match, and the jigsaw continues for cubic miles. These rocks crystallised from liquid melt, and as they solidified the locks between crystals were set for millennia.

On the geological map the intrusions appear as angry blotches. It is immediately obvious how their outcrop disdains that of the surrounding rocks, cutting across other geological boundaries. What they cannot push aside they can incorporate, as a stockpot may swallow a parsnip, and granites may include darker lumps of other rocks that have been included in this way. Closer study has revealed that there are differences between granites; some are finer grained, or have different mineral or chemical composition. Hence it becomes clear that Ben Nevis (Devonian) has formed in several stages. Wind and rain and ice do not miss such subtleties when seeking where to erode.

Granitic soils are thin and poor. One might say that the main crop of the

---

*I should add that I use 'granite' in its broadest sense. There are several other types of coarse-grained igneous rocks distinguished by their mineral composition, such as syenite, some of which are known from these Scottish intrusions, but all are lumped together here.

Cairngorms is skiing. Ice lingers there. And it has ever been so back into the Pleistocene. Members of the heath family (Ericaceae) can tolerate these paltry peaty composts, not just ling, but berry-bearing cranberries and crowberries. These little fruits make good pies and passable jam. They are rich in vitamin C: in fact, they have prevented scurvy. In areas covered by the Scandinavian or Newfoundland Caledonides, and still older areas of Nordic Precambrian, these humble berries have been the only source of this important vitamin in winter months. The berries were stored beneath water, in which state they keep very well for months. Children would pop a few into their mouths and swallow them like pills because they are bitter.

Granite resists the importunities of ice better than many rocks, but has to succumb in the end. Boulders of the granite itself have been pressed beneath tons of ice in the Pleistocene to scour the bedrocks; the rock, one might say, has been used to scratch its own face. But once the ice vanished between seven and ten thousand years ago its imprint was left unmodified by time. Glen Clova is scoured by huge corrie cliffs where glacier ice was once seated. Snow lingers in these to this day after it has melted elsewhere.

What the Highland and North-west Irish granites intrude, nearly everywhere, are gneisses and schists: metamorphosed sediments, for the most part, more rarely metamorphosed igneous rocks. Gneisses are broadly banded rocks, often with a pink and black appearance – the colours of old-fashioned breeds of pigs. Schists are more finely sheeted rocks – like a croissant, perhaps, and occasionally just as flaky. They always glisten from the mica they contain, dark or light in colour according to their mineralogy. Schists often started life as shales or mudstones, before the heat and pressure got to work upon them, but there is little trace of their origin now. The way they break – following their *foliation* – often has nothing to do with the bedding of the sea floor where they once originated, and everything to do with the minerals that have grown anew under the influence of metamorphism. The banding of gneisses and schists can wander any way – vertical, horizontal, ruckled like a crazy Italian ice cream, folding back on itself – it seems as if chaos rules in metamorphic terrain. There is a plasticity about it all that apparently defies rational interpretation, like trying to make sense of a plate of tagliatelle. But there is a hidden order there.

One key to unscrambling some sense from these madly distorted rocks proved to be hidden in metamorphosed sandstones (psammites). These do not so readily lose the original structures acquired when they were sediments. Some of these structures even endure high grades of metamorphic oppression. These structures tell us *which way up* the beds are. This may not seem like much (nor would it be, in southern England), but in the Scottish Highlands it quite literally turned previous ideas upside down.

One example will illustrate the nature of the problem. When some kinds of coarse marine sandstones accumulated they did so in a kind of slurry, with all the grains churning around in suspension in the sea water. As the current carrying the grains lost energy the larger grains, as one might expect, fell to the sea floor first, followed by finer and finer grains. The result is a bed – often a few tens of centimetres thick – which is *graded*. The bottom of the bed can always be identified by the coarser material. Such turbidite rocks frequently occur stacked one upon the other for hundreds of metres. These are the kind of obvious structures which may survive metamorphism. Now the rub is that, in the Caledonides of the Highlands, where the right kind of rocks came to be carefully examined many proved to be completely upside down! They were overturned as effectively as you might overturn a book so that the last page lay at the top, the first page at the bottom. There was only one conclusion possible: the whole slab of country containing these rocks had also been inverted.

Thus it was that some of the dragons of ignorance came to be slain. One of the chief slayers was E.B. Bailey, a great geologist by any standard. He breakfasted on porridge and refused luncheon, so that neither his mind nor his legs should be diverted from the rocks. *Any* dragon would have found him formidable. He appreciated how the Highlands could be understood by analogy with the Alps – and when he wrote in the 1930s this was long before the concept of Iapetus and the common *explanation* of both Caledonides and alpine structures. What he saw – hidden in the Highlands, but blatant in the Alps – was that much could be explained by *nappes*. Nappes are related to the great Moine Thrust described in the last chapter. They are vast, slithered masses of the earth, squeezed out from a 'root zone'. They are seated on great slides. One way to imagine a nappe is to take a rolled-out sheet of pastry, and pluck out a handful from the centre; pull on it horizontally until it attenuates and then detaches. The gob of pastry you hold in your hand is a nappe: part of it is turned upside down. The 'root' zone is where you plucked it out.

But on the ground – in the reality of Scotland – you have to imagine this process drawn out through tens of miles. A whole mountain may be represented by the dabs where your fingers squeezed the pastry. So a Cnoc or a Beinn may be composed of strata that are turned upside down, and even these major folds folded yet again, so that the little contortions you see in the cliff – the details – are like grace notes upon a great theme. Even the merest wiggle in the schistosity may be a reflection in microcosm of the huge design. Truly, this is a case where the eddy represents the waterfall, the crenellated parapet the castle. Those that celebrate the human imagination that can swim among galaxies, or dissect the atom, seldom wonder at the genius that can read a mountain range from the merest fold, or infer the vanished past from a hint seen only as a modest bluff among the bog. These scientists are heroes of perception, being translators of the magnificent from the mundane. Edward Bailey, Janet Watson and Robert Shackleton had this knack of seeing the whole picture from the bruised corner of a partial sketch. The highest part of the nappes, where the strata are 'right way up', which can be seen to advantage in the Alps, are in Scotland deeply eroded away, so that in many places only the lower, inverted limb remains. This is where those overturned turbidites come in to the story. From the evidence of inversion in a few beds which have escaped metamorphic obliteration along places like Loch Tay comes an awareness of the topsy-turvy nature of a cubic mile, and thence a vision of a colossal overfold affecting hilltop and boggy glen alike.

But what of depth? How can we find proof of the incarceration of these mighty folds deep within the Caledonian fold-belt?

The traces of this history also remain, locked in the minerals. All these metamorphic rocks are time capsules, set at least 350 million years ago, and still preserving the story of their genesis. From inferring structures of mighty magnitude from a mere mountain we now seek the story concealed within a single crystal. Certain minerals will only grow under conditions of temperature and pressure found deep within a mountain range: find those minerals and it is almost like having a thermometer and a pressure gauge in your hands. You can experiment in the laboratory to reproduce the right conditions to grow the right minerals.

A few of these minerals bear familiar names – garnet, for example. Garnet-mica-schists cover great areas of the Highlands. In truth, the garnets are usually disappointing compared with the plum red, semi-precious stones popular with Victorian jewellers. They are lowly lumps dotted over the schist surface, like nuts in nougat. They often have a spherical appearance, more like a hazel-nut than an almond, and the crystal faces are not particularly clear; but good specimens of the red variety almandine with twelve or twenty faces are not uncommon. Almandine garnet will form only at a high pressure of 12 kilobars and at temperatures of about 900 celsius and hence its presence is an indicator of where and how this part of the Caledonides was deformed within the ancient range. Two other famous 'indicator' minerals which switch over at particularly high temperature and pressure are kyanite and sillimanite: whole tracts of land are fingerprinted by their occurrence. Yet further minerals shuffle their elemental composition around as temperatures and pressures change – a kind of crystallographic thermometer. These minerals are studied by cutting sections through the rock, and grinding and grinding until the whole becomes transparent. Mounted on a microscope slide for study under a light microscope the minerals betray their identities by colour, shape, and their behaviour under polarised light. These days, individual mineral grains can be probed with electrons to reveal their elemental composition directly; as with so many things scientific, technology allows us to go smaller and smaller with greater and greater accuracy. Just recently, new techniques have allowed the investigation of individual inclusions *within* single crystals. Curiouser and tinier in pursuit of the bigger picture; it is a paradox.

Rocks are composed of combinations of minerals. One might compare them with a *pointilliste* canvas, in which primary colours serve to make infinite combinations of shades and shapes. Minerals, like colours, are finite in number, yet no two rocks are completely identical. They can be painted coarsely or finely; the minerals can be used in various combinations; the texture can vary.

A few minerals account for much of the Highlands: quartz, felspar, mica, hornblende* more than most. Both granites and gneisses can be built from these alone. Their abundances vary wildly from rock to rock. The pinkness of many granites and gneisses comes from the felspars they contain; others are milky white. Often the felspar crystals appear without clear shape, but in granites, for example the granites from which the dour, eternal houses of Aberdeen are constructed, big crystals as long as your thumb are common. The hornblende gives the dark colour to many schists. The glitter of all these rocks comes from the mica. The planes that reflect the light are flat crystal surfaces, and these in turn are but the mirror in macrocosm of the way the atoms composing

---

*felspar, mica and hornblende are actually families of minerals with a variety of chemical compositions, but sharing a common crystal structure. The dark mica, biotite, is common in some schists, for example, while the light-coloured mica, muscovite, dominates in others, affecting the whole appearance of the rock.

Striped metamorphic rocks typical of the Scottish Caledonides: horneblende schists and quartzites forming the Falls of Lochay, Killin, Perthshire.

the minerals are stacked, in microcosm, in countless thin sheets. In gneisses and schists the micas are aligned with the foliation, which is why they glint when the rocks are turned in the light; the light falls on their parallel crystal surfaces all at once. Mica crystals are often small and seen with difficulty with the naked eye, but, rarely, in veins of a rock known as pegmatite, the crystals have grown more slowly, and then they can be very large. You can flake them with a penknife into a myriad tiny sheets, like mineralogical dandruff.

If the Highland rocks are a recipe of minerals and textures, the minerals themselves are a recipe of elements. And those elements comprising the common minerals are surprisingly few: silicon is the prime ingredient. What carbon is to life, silicon is to rocks. Its molecules join with oxygen and with other elements in prodigious variety. To silicon and oxygen add aluminium, sodium, potassium, iron, with various amounts of hydroxyl (OH), and there is enough to brew mountains of granite, moors of gneiss, glens of schist. What is seen in the landscape is the end product of the subtleties of mineralogical recipes concocted in the depths of the Caledonides and then ground down by the millennia of erosion, traumatised again by ice, and then reconstituted finally by the hand of humankind.

This brings up one of the questions which can be asked especially of wild landscapes, like those of the Highlands. Its openness, and the almost ubiquitous presence of rock, hinted at just below the surface even when not actually visible, suggests a primeval quality, as if here were the one true wilderness. But is this correct? Or is this apparent wilderness in its own way as manufactured a landscape as the neat patchwork of fields and hedges of the ancient countryside of England? The effects of the Clearances are well known – the forcible removal in the late eighteenth century of agricultural peasantry from their modest living with the introduction of sheep farming (1782–1854). This was partly a result of the breeding of Blackface sheep, which prosper on moorland, and which were imposed with notorious singlemindedness by the landed gentry at the expense of the human residents, not to mention the highland cattle: that sad story of depopulation is a familiar one. Its contribution to the present wildness was bought too dearly, for all that the world benefited from the Scottish diaspora, and the scattering abroad of native talent.

But at a deeper level, there is a question whether moorland itself is 'natural'. Tundra-like, but extending further south than it should by comparison with the rest of the northern hemisphere, more than half the Highlands is covered with it. It is one of the wonders of our islands, and one of the reasons why foreign naturalists come to Scotland. Its defining characteristic is an underfelt of *peat*, the accumulation of plants over hundreds or thousands of years in the wet ground. The poor drainage of Caledonia undoubtedly accounts for its widespread formation just here; black or brown and more or less dense, the roots of the ling or grasses of which it is composed can usually be seen if it is crumbled between the fingers. Cut into bricks and dried it still provides fuel for crofters. A peat fire glows dully red, but makes real warmth. What is burning is the undecayed carbon, just as in coal, which survives because of the waterlogged conditions under which it accumulated. Through Ross and Cromarty, Sutherland and Caithness great swathes of blanket bogs cover even the hillsides, frustrating geologists who wish to get at the real rocks, which they can often only see in the bottom of the gullies or hags cutting through the peat. The peat itself includes pollen from the plants that were living while it accumulated. Pollen experts can identify the species, and hence open another window into the past to look at former vegetation.

Stumps of trees are quite commonly found beneath the blanket peats, and hence the question whether the Highlands were once all Caledonian wildwood. Was early man the culprit, cutting and burning, and turning all into wilderness

that we now, erroneously, think of as natural? Rowan and birch nuzzle in protected bluffs – but were they once ubiquitous? This argument is of more than
academic interest, because those who wish to plant spruces in massive patches
might argue that what they are doing is only, after all, returning the landscape
to what it once was. Clearly, there *were* once more trees in more places. The fossil
pollen indicates more trees at the time when wildwood prospered (about 4500
B.C.), birch predominant in the North-west Highlands, Scots pine centrally, and
oak/hazel woodland further south. The north-west was probably always largely
tundra. Climate changes dictated what happened to the wildwood as much as
man, all acting upon the bedrock reality of the nutrient-poor rocks of the
eroded Iapetus Alps. Still, it is consoling to find that Scots pine would have
dominated a large tract of land. This tree embodies spareness and wilderness;
the angularity of exposed specimens seems to belong to a Japanese landscape,
as if it might have been figured in a Hokusai print.

But the moor is invasive, and once established appears to inhibit the growth of
any trees. The Forestry Commission and others have to till deeply to re-establish
the right conditions for their growth; peat and nutrient poverty prevail. Rain and
exposure and leaching dictate that forest tends to revert to moor over hundreds of
years. In dry periods, fire, too, would have contributed to loss of forest, although
modern historians apparently do not favour Viking rampage as a prime cause of
deforestation. So it is that the metamorphic Caledonides provide the most extensive stretches of moorland in the British Isles. There is yet more in Caledonian
Wales, of course, and in geologically younger Dartmoor and Exmoor. But Scotland
alone has enough to support golden eagles and pharalopes, and red deer. Deer
grass, cotton grass, reeds and sedges, ling in patches; mosses and clubmosses and
liverworts and lichens, which require little in the way of nourishment but much
in pure atmosphere, these provide the common ground. Clubmosses are living
representatives of a group of plants that made trees in the Carboniferous, and
now prosper where little else can earn a living. There are ferns, too, as there
always are when there is moisture everywhere. The Royal Fern, *Osmunda*, is the
most stately of these: to the casual eye it scarcely looks ferny at all, but it, too, has
fossil forbears. The insectivorous plants such as the sundew, *Drosera*, have found
another stratagem to get food. It is odd how primitive survivors ally with weird
specialists like *Drosera* in this inhospitable moorland. It is a tough battle for life,
there, and perhaps the old campaigners and those with the latest gizmos are the
ones that survive.

Sphagnum moss grows where land is permanently soaking, its star-shaped rosettes of leaves looking like so many minute, leafy chrysanthemums. It forms pink or green cushions decimetres across that often look luxurious and inviting to lie on. Woe betide anyone that tries, as *Sphagnum* often perches on top of quaking bog. It is amazingly absorbent stuff: it can be wrung out like a soaking towel. It grows where nothing else can get a foothold, and its old stems make dense, black peat which gardeners love to mix into their composts. There are a whole batch of little mushrooms, species of *Tephrocybe*, *Omphalina*, and *Galerina*, that will only grow in *Sphagnum*; most of them are orange to olive coloured and they have long stems rooted deep in the bog. The *Sphagnum* bog is a distinctive little ecosystem all of its own.

At the top of the mountains, imposed by altitude as much as rock, the alpine flora is different again. Some of the plants have relict distributions; stranded after the Ice Age, their nearest neighbours may be living in Norway or Switzerland. While many of these plants are humble little herbs, others, like the mossy campion, (*Silene acaulis*) are delightful – with tiny, bright pink flowers dotted on a delicate pincushion of leaves. Ben Lawers is a famous peak with an exceptionally rich flora, which has been both protected, and opened up to the public with a waymarked route to make the walk to the summit less arduous. The rocks which make up Ben Lawers are more calcareous than are the usual highly acid Highland metamorphics, and this partly accounts for the fact that some species of plants growing there are found hardly anywhere else in the British Isles. It is a delight to have truly alpine flora within our shores, and how much better these flowers look growing where they should than sitting sadly on a suburban rockery.

Along the flanks of Ben Lawers huddle the remains of low buildings – shielings – which provided shelter from bad weather for man and beast. These buildings have long since fallen into disuse and decay, as have others in the Highlands as farming declined. Like the stone walls of abandoned fields, these simple buildings have been thrown up from the loose rocks that lie around the fields and in the beds of streams, and perhaps roofed originally with turf. Highland crofts were often simple affairs constructed from local stones, of which there is no shortage. Granite boulders prove particularly durable. Traditional long houses catered for human habitation at one end and animals at the other. There was no question of defying the weather. Buildings are tucked into sheltered positions in *straths* and along burns, and the old inns must have seemed welcoming indeed to the

winter traveller. Whitewash is about the limit of decoration. But such vernacular dwellings do seem entirely happy within the landscape; grown from it, they
remain part of it. Whether you feel the same about some of the hotels and
castles constructed (often not until the nineteenth century) in Scottish baronial
style depends on whether you find them romantic or grandiose. Beturreted and
begabled, they were built to last, from sandstone or granite, like some fantasy
of Mad King Ludwig of Bavaria, or the ideal wicked witch's castle as conceived
by Walt Disney. The pattern obviously had international appeal. I have seen
examples across Canada: huge hotels constructed in Quebec City in the east
and Jasper in the west. They always look somewhat improbable, as if waiting
for a prince in doublet and hose to march out of the front portal or a maiden
to loosen her locks from the top of one of the turrets.

If the Highland Boundary Fault is followed westwards it crosses on to the
Isle of Arran. It would be possible to include Arran in several places in this
book, because it has such a diverse geology, and the excuse for doing so here
is only because the north of the island includes a sliver of the metamorphic
Highlands. If the British Isles are geologically rich in relation to their size, Arran
is richness distilled. What the first cadaver is to the medical student, Arran is to
the fledgling geologist. Student parties flock there. There are sedimentary rocks,
with and without fossils, igneous rocks, metamorphic rocks, and a scattering of
glacial phenomena. I, too, crawled over the carcass of the island muttering my
litany of names of igneous rocks and wondering whether man was or was not a
taxonomic animal, concluding that if he was, he had undoubtedly invented too
many labels to learn. Arran is oval, its long axis twenty miles north to south.
It is a jewel of a place, the more so because you reach it via a ferry from
Ardrossan on one of the dullest stretches of coastline I know. Arriving there is
like an emigration. It rises steeply from the coast, so you are never more than a
minute or two away from a good walk. It has been a popular holiday resort for
the better-off Glaswegian for a century, and the consequent solid hotels provide
breakfasts with white pudding as well as black, and serve that undistinguished
beer appropriately known as 'heavy'.

There is an extraordinary difference between the eastern side of the island
overlooking the Firth of Clyde, and the west. Brodick Castle Gardens on
the eastern side are a subtropical plantsman's Mecca, crammed with exotic
rhododendrons and azaleas, gigantic ferns, branches carrying epiphytes, and
great specimen trees. The wonderful blue Himalayan poppy (*Meconopsis*) – with

Typical scenery in the Grampian Highlands. Tyndrum lead mine in the foreground, Ben Lui formed of mica schists (metamorphic) in the background.

a flower the colour of the Mediterranean sky – seems to grow there like a weed. Dedicated gardeners in the south-east of England have been known to kneel and pray before this choosy species to induce it to flourish, but, in truth, it only wants to grow on the east side of Arran. Brodick Castle is sheltered from the prevailing winds by the heights of Goat Fell, so it is a haven with a warm, moist climate mollified beyond its latitude by the North Atlantic Drift. Inverewe, much further north again, is another such favoured and protected garden. Here in western Scotland rhododendrons find a climate matching that of their own origins in the Himalayan foothills.

But the ocean-facing west side of Arran is bleak, Highland-style country, with deer grass and sheep, and lopsided trees that lean away from the ocean as if cringing in terror. At Lochranza in the north-west of the island toughness is the most obvious aspect of the local buildings and a keep-like castle adds the last touch to the wild aspect. Thus local climate works upon the flora on its geological foundation to blast it or favour it according to its whim.

71

The metamorphic rocks to the north of the Highland Boundary Fault on Arran are part of the Dalradian Series. This group of rocks forms a great belt south of the Moine Schists and north of the Fault, running from Kintyre and Loch Awe in the south-west along the Caledonian trend to Portsoy in the north-east. Dalradian rocks are, in general, younger than the Moines, being latest Precambrian to Cambrian, and are less metamorphosed in places. They continue south-westwards into Ireland. In South-west Scotland, the Dalradian is taken up in one of the great alpine structures, the Iltay Nappe. Slabs of country around Loch Fyne are inverted as a result. To the North-east, Dalradian country is intruded by the huge and implacable late Caledonian granites of the Cairngorms, Braemar, Aberdeen and Mounth — Munro country, because the height of the mountains means that they enter the list of desirable peaks for the mountain climber. In Arran, the local Dalradian is called the Schistose Grit, comprising some rather indifferent flaggy beds. The granite forming Goat Fell on Arran is not one of the Caledonian ones, but intrudes the Schistose Grit as part of the much younger, Tertiary igneous activity, associated with the islands of Rhum, Skye and Ardnamurchan. The gradients and heights on Arran are almost invariably related to the hardness of the rocks, and Goat Fell is the hardest and steepest. There are wonderful views all the way up the steep climb. I wish I could report that the view from the top makes the effort of climbing worth it, but on the occasion I reached the summit one of those impenetrable Scotch mists came down, so I spent the time stumbling and cursing.

I remember seeing contact metamorphism for the first time, where the intrusion of Goat Fell had baked the country rock it had intruded — proof that the intrusion had risen from deep in the Earth, and was hot. I recall Judd's Dykes, where the vertical insinuation of dark volcanic rock cut through the country like a black blade, and where dyke cut dyke to give a relative chronology of intrusion. There were chilled edges on these dykes, too, where they had cooled to glass as the hot magma encountered the cold slate. The hidden landscape became revealed: or, rather, I learned to see what had always been there.

The one dragon that is alleged to lurk even now in the Highlands also has geological connections: the Loch Ness Monster. Its myriad interpretations include the common element that it is a survivor, a messenger from the Age of Dinosaurs, a shy colossus. Loch Ness itself is a curiosity enough. Long and narrow, it lines up with Loch Oich, and Loch Lochy and the sea Loch Linnhe to point a line on the map that you can see from the Moon. The lochs sit in the Great Glen, which cuts the Highlands in two and joins the Moray Firth with the Firth of Lorne, such that a sea level rise of even modest proportions would float off the North-west Highlands as a Gaelic island. It is an interruption in the Caledonides, but its direction obviously follows the Caledonian trend. The dark depths of the Loch are usually placid, but can be whipped into urgent waves. There are mysteries here without invoking dragons. The Great Glen follows a huge crack in the earth, a vast fault, cutting to inconceivable depths. It is a San Andreas of the British Isles, sunk in inertia now. But once earthquakes shook the ground around it, and vast tremors troubled the Earth's crust, and wrenched the ground to the north one way and the ground to the south another. For this was not a fault that simply let down one side relative to the other – rather, in Devonian times, the two sides slid (or laboriously squeezed) past one another, so that the area north of the great Glen Fault previously lay relatively further to the east. This kind of displacement has been recognised for some time, but just how much movement was involved has been controversial. The San Andreas Fault has thousands of kilometres of movement upon it, and one research worker suggested as much as 2000 km along the Great Glen; but the lowest estimate is probably 80 km, and present consensus is about 200–300 km. It is not very easy to produce a definitive answer because most of the arguments depend on 'matching' geological features now displaced on either side of the fault, and, anyway, movement may have been protracted. Even now there are tiny trembles. Its heyday was doubtless in the later Devonian history of the Caledonides, after Iapetus had closed, but while the crust still had to accommodate stress.

Did the strange shape of Loch Ness breed the notion of monsters, or was a late rumbling on the Great Glen Fault a progenitor of dragons? No one would be more delighted than a palaeontologist to discover some plesiosaur, stranded by time, still fishing and breeding in Loch Ness. Think of the joy of discovering the living coelacanth fish, *Latimeria*! Sadly, 'Nessie' seems to be no more than silly season fodder, a wheeze by the Scottish Tourist Board to encourage visitors, for it is a vertebrate that leaves no bones, a monster that cannot be dredged, a shape without a fixed design – a dragon of dreams, half seen across the loch among mist spilled down from the hills. When dusk comes, Loch Ness can be a spooky place, and then the exhalations of the Earth solidify in dreadful forms, and the deep, dead fault might seem to convulse once more. If there were really a monster, Loch Ness would still be the best place to hide it.

6

The Southern Uplands

IT IS NOW necessary to take a jump across the Midland Valley of Scotland to the Southern Uplands, where once again, the rocks belong to the Caledonides. The northern edge of the Southern Uplands is defined by another great fault, the Southern Uplands Fault, running more or less parallel to the Highland Boundary Fault, from Ballantrae in the west to Dunbar in the east, 130 miles. The ground between the two faults has been let down to comprise the Midland Valley, a broad region now occupied by rocks mostly younger than Caledonian, and utterly different in character. The great, threefold division of Scotland – Highlands, Midland Valley, Southern Uplands – is first a geological one. It affects just about everything else: cities, industry, intensive farming, universities are all in the Valley. Sheep, and grouse, and heroes worthy of Sir Walter Scott, lie to the north and south.

As happened to the north of the Highland Boundary fault, so it is to the south of the Southern Uplands Fault: a resurgence of wildness, bare moorland, poor soils, tumbling burns. And, again, the Caledonian trend can be traced across the Irish Sea into Northern Ireland, in counties Down and Armagh. This tract of Ireland is a geological continuation of the Southern Uplands; the rocks do not see physical separation. But there is a great difference between the Caledonides in the Southern Uplands and in the Highlands, because the rocks are not strongly metamorphosed in the former. There are good, sedimentary rocks in the Southern Uplands, with fossils to be discovered in places. It is known that this great tract of land was laid down beneath a deep sea during the Ordovician and Silurian, and fossils have left their signature for both time and place. The rocks comprise predominantly dark shales, paler grits, grey or greenish mudstones, conglomerates here and there, repeated time and again, for mile upon mile. There are some dark volcanic rocks near Ballantrae, while granite intrusions make up the mountains of Cairnsmore of Fleet, Criffel, and around Loch Doon. There is much tough and remote country (hardly the Lowlands of the guidebooks), for all that the heights do not often approach those in the Highlands.

One of the holy places in geology is at Siccar Point, a few miles west of St Abb's Head on the Berwick coast. A short walk from the main road will take you there. This, one might even say, is the type example of *unconformity*. Here James Hutton (1726–97) observed nearly horizontal Devonian sandstones resting upon Silurian slates and grits, which are vertical. Because those Silurian rocks themselves were once laid down – horizontally, naturally – upon the sea floor, here was proof that they had to be uplifted, and turned vertically in a great convulsion of the earth, and then eroded deeply, before the deposition of the Devonian upon them. The contact was an unconformity. Geological processes must have required time, immense stretches of time, to accomplish all this. John Playfair visited Siccar Point with Hutton in 1788, and described the inspirational effect of the place in a famous passage:

> The palpable evidence presented to us, of one of the most extraordinary and important facts in the natural history of the earth, gave reality and substance to those theoretical speculations, which, however probable, had never till now been directly authenticated by the testimony of the senses. We often said to ourselves, what clearer evidence could be had of the different formation of these rocks, and of the long interval which separated their formation, had we actually seen them emerging from the bottom of the deep? We felt ourselves necessarily carried back to the time when the schistus on which we stood was yet at the bottom of the sea, and when the sandstone before us was only beginning to be deposited, in the shape of sand or mud, from the waters of a superincombent ocean. An epoch still more remote presented itself, when even the most ancient of these rocks, instead of standing upright in vertical beds, lay in horizontal planes at the bottom of the sea, and was not yet disturbed by that immeasurable force which has burst asunder the solid pavement of the globe . . . The mind seemed to grow giddy by looking so far into the abyss of time . . .

In such richly coloured language were great cycles in the history of the Earth consecrated. The Caledonian folding of Silurian rocks predated Devonian sediments, and thus were the earth movements timed. One slightly shudders to think in what magisterial concatenations of adjectives Iapetus would have been celebrated, if Playfair had only known about it. These famous visitors to Siccar Point could not have anticipated that what 'burst asunder the solid pavement of the globe' was the movement of great slabs of the crust itself – indeed, they might have found such explanations overly catastrophic for their taste.

The Southern Uplands is the margin of Laurentian Iapetus: go further south again and the Lake District lay on the opposite side of that great and vanished ocean in the Ordovician. The fossils that date the rocks are mostly graptolites: an extinct class of floating, colonial animals, whose matted remains often cover the bedding surfaces of black shales with whitish, saw-edged lines, as if they had been scribbled by an abstract artist (their name refers to their resemblance to writing on the rock). Graptolites are wonderful fossils for dating rocks, because they changed rapidly through geological time, and because particular species floated widely around the world. Thanks to them it was recognised that the Uplands rocks are progressively younger southwards: Ordovician in the north, Silurian in the South.

The use of graptolites to unscramble what is, in detail, astonishingly complicated ground, can be largely credited to Charles Lapworth, who has already been encountered as one of the geologists who 'solved' the Moine Thrust. One of the most famous rock sections in geology is near Moffat, in the heart of the Southern Uplands. The section is exposed along Dob's Linn, a name which is as familiar to geologists in Beijing as to palaeontologists in Alma-Ata. It is, in truth, a bleak place. There are precious few trees on the steep-sided gulleys; on a windy winter day the valley funnels every chill towards you, and the geological hammer weighs heavily in numb fingers. Persist in splitting the black shales, though, and you will be rewarded with a picture gallery of graptolites, showing white specimens laid out on flat bedding surfaces as if they had truly been written on the rocks specifically to elucidate history. Much of the Ordovician and the early Silurian can be collected in sequence, species by species, and what a joy it must have been to Lapworth when he realised that these uncomfortable grassy slopes could unlock the history of a slab of our islands which had hitherto been arcane. The plain cottage in which he lived while he worked as a schoolmaster, and, as an amateur, transformed our understanding, lies nearby on the main Moffat road, marked with an appropriate plaque.

The great tract of Ordovician and Silurian shales and grits produces a rather impenetrable countryside of steep-sided, bare hills, dissected by burns, which is

The famous unconformity on the Berwick coast between the upper Old Red Sandstone (Devonian) and the folded Silurian rocks beneath, where Hutton and subsequent observers understood the cyclical convulsions of the Earth.

The Devil's Beef Tub, Dumfries, excavated by glacier ice in the Silurian shales and greywackes of the Southern Uplands.

*Graptolites, 'writing in the rocks', the fossils of extinct colonial organisms which helped to date the Ordovician and Silurian rocks of the British Isles. (*Didymograptus*, Dyfed, S. Wales).*

typical of much of the Southern Uplands. The rocks lack contrasting character which might produce something more spectacular, although the dark hole known as the Devil's Beef Tub north of Moffat is grimly impressive. One can see what a barrier the area must have been between the old kingdoms of England and Scotland.

Although Lapworth published his seminal article in 1878, Dob's Linn was not simply filed away under the heading: 'problem solved'. It has been visited and revisited. Lapworth collected foot by foot, but now my colleague Henry Williams has collected inch by inch, and found new graptolites. Recently (1985), this already famous place has become *the* reference section for the base of the Silurian System worldwide. My only regret is that trilobites are so very rare there. The Ordovician sea floor may well have been lacking in essential oxygen for them to breathe, which did not worry the graptolites floating far above in the open sea. The dark graptolite shales were accumulating offshore, and trilobites mostly preferred to live nearer to the edge of the continent. To find them, you have to go north and west to the district around Girvan (Ayrshire), where the scenery as well as the fossils is more varied. Here you can find conglomerates that tipped off the edge of the Ordovician Laurentian continent, and, in places, whole trilobites whose closest relatives are known from Cincinnati. Mrs Eliza Gray spent a lifetime collecting them, and eventually her collection was given to the Natural History Museum.

But the recognition of Iapetus has had as profound effect upon the way we see the Southern Uplands as it has upon the rest of northern Britain. This, one might say, is the great advance since Lapworth: the Southern Uplands are the 'scrapings' of Iapetus. As Iapetus closed, the subduction zone down which the oceanic crust plunged to destruction lay along the Solway Line to the south (see p. 39). The sediments that had accumulated off the edge of Laurentia were piled up against the Laurentian margin in a stack of slices: this is why the whole of the Southern Uplands is typified by so many repetitions of sequences of rocks of similar age, and also explains rather neatly why the rocks become progressively younger southwards. This arrangement has been seen in comparatively recent subduction sites, as in the Makran region of Pakistan, where it has been called an 'accretionary prism'. Lapworth himself had explained the repetitions by endless pleating and folding of the rocks (and there is undoubtedly a lot of folding, too), but the idea of these oceanic sediments scraped up at the edge of Iapetus has a tremendous simplicity that explains so much at a single move. One thinks again of the Moine Thrust, and how a shift in perspective changed not only the focus, but the substance of what was seen.

The Scotland that I have described is a geological entity, the great pile of the Caledonides, lying at the edge of a Laurentia more ancient still. There is no escaping the fact that the political boundary between England and Scotland coincides with a fundamental geological one. Those with a taste for the meta-physical might like to make something of this, a kind of geological control on character and history. Much as I like simple explanations – after all, they are so much easier to remember – this one gets only muted support. On the positive side, there is the hardiness needed to survive poor soil, a cool climate, and, for centuries, difficult communications outside the central valley, which will almost inevitably produce an invincibility of spirit and a tendency to clannishness. And if there were Caledonian genes they would have been stirred around in small pots. In historical times, the Highlands provided a refuge as new peoples moved into Britain from south and east: the face of the land, and ultimately its geology, was responsible for this haven. But that is as far as it goes. No doubt a case could be made for whisky as a geological phenomenon, because mysterious qualities of the water – Spey or Islay – are often invoked to account for the distinctiveness of this species or that, but the most important ingredient is probably the skill of the manufacturer. But it is true that a people hardened in the Highlands are capable of living in most places. When I visited the hottest and toughest part of central Australia, the cattle barons at the edge of the Simpson Desert carried Scottish names, and the stations were called Tobermory and Glenormiston. While rain and wind battered the originals, sun and yet more sun bleached their antipodean homonyms. In both, doubtless, a weatherbeaten Scotsman poured himself a whisky and cursed the elements.

There is another patch of Caledonian country which lay on the north side of Iapetus. This is in western Ireland, in the counties of Galway and Mayo, west of Lough Mask. Were it not for a covering blanket of younger, mostly Carboniferous, rocks no doubt it would be seen to extend further to the east, following the Caledonian trend. It is a wonderful stretch of varied country, and the variety is partly because the Ordovician rocks include almost every shade and nuance of sedimentary or volcanic rock. The rocks accumulated in a great trough, interrupted by volcanic eruptions, near the edge of the Laurentia. The harder volcanics often form ribs running up the hillsides, while streams seek out the softer shales. Because the peasant farming economy continued so much more recently here than in Scotland there are rough walls and green meadows everywhere. Add to this a scattering of small loughs, and long inlets or bays like Killary Harbour – which are *rias*, valleys 'drowned' by the sea level rise which followed the Ice Ages – and tiny roads which meander specially to present the best view of the geology, and you have one of the most delightful areas in the world in which to try to find a trilobite. The town of Westport has bars that seem never to close, and the entertainment in the bars is animated conversation, not the clatter of disco. And after summer rains the fields erupt with white wild mushrooms. No dragons here, or else they would be ones that could be led with a halter.

7

The Land of the Ordovices

IN NORTH WALES, south of the town of Porthmadoc, the main road crosses a wide estuary on a causeway; to cross it you must pay at a toll bridge. If you look northwards from the toll booth you see a small hill where quarrying is in progress. This is the hill of Y Garth, and on its far side there are trilobites.

To reach the old quarries where the trilobites can be found (the functioning quarries are in igneous rocks, and so lack all fossils), find a path that takes you over the top of the hill towards meadows that line the estuary. It is an old path, walled with the local geology on both sides, paved with cobbles, and carefully made. Overgrown walls, almost ubiquitous in Wales, are a testament to more populous times, when the cobbles would have rung with the clatter of the hoofs of packhorses, and there would have been fields where trees now gradually nudge stone walls into oblivion – walls which once provided field boundaries. There are trees, also, in the old quarries, but there are still plenty of slabs of shale lying about. Split these slabs and the chances are that in ten minutes or so the first trilobite will be found. It will be *Angelina sedgwickii* Salter, 1859. It is also likely that it will be a whole specimen; this locality is unusual because complete trilobites are so common. Usually, finding a complete trilobite specimen is a cause for celebration. What is absolutely certain is that the fossil will be squeezed, stretched, twisted, contorted, and changed from its original appearance. On Y Garth the rocks have been distorted, and the trilobites within the rocks have been treated like passive victims on a rack. What you see on the surface of a piece of shale is a metaphor for what has happened to the whole slab of country. And that, too, is but a tiny part of the southern side of the vanished Iapetus ocean. So the suffering of *Angelina* is nothing less than the writhing of the Earth's crust in the great compressive forces unleashed as Iapetus closed, when rocks were squeezed between the opposing sides of Wales and Laurentia as in a gigantic vice. Its pain records the birth pangs of our own islands.

These North Welsh rocks are not strongly metamorphosed like those in the Highlands because they were not heated as much; there are no gneisses and schists in the Welsh Caledonides. But the softer rocks, like the shales carrying the trilobites, have plastically deformed in many places. The more robust rocks – sandstones, conglomerates and limestones and the like – have often escaped the worst, while shales accommodated the stress. The good news is that fossils have survived in many places, so that we can read the history of life from the rocks.

The rocks exposed over much of Wales are part of the southern limb of the Caledonides: Snowdonia, the Harlech Dome, mid-Wales, the coast at St David's. Wales is the type area for the first three of the geological systems which routinely include the traces of shells of extinct animals: Cambrian, Ordovician and Silurian, in ascending order (550–412 my). The description of time itself is rooted there. 'Cambria' is, of course, Roman Wales, and the name survives in the Cambrian mountains; the modern Welsh Cwmrw is from the same root. The Ordovices were the last of the ancient tribes to submit to Roman domination, holding out from their strongholds in North Wales; the Silures were another barbarian tribe from South Wales. They provided appropriate labels for naming rocks, names which still have a pleasing euphony. The systems were applied, with, it must be said, a great deal of bickering, during the course of the last century, and now are understood around the world. Whether a geologist is Chinese or Latvian he knows what Ordovician means and thereby acknowledges the Welsh hills in his understanding of time itself, and a link with the land of the Ordovices.

To walk in Caledonian Wales is to cover hallowed ground, where chunks of geological time were recognised for what they were. Here trilobites were used to see the succession of the rocks and to map the ground for the first time. Great men like Adam Sedgwick and Sir Roderick Murchison sought to imprint their will upon chaotic Nature, wrestling with the geology of mountains still obscure, whose previous celebration had been in the bardic verse of Tam y'r Nant or Goronwy Owen. They saw volcanoes long extinct; they found faults nearly as profound as that forming the Great Glen; they discovered a hundred extinct species pleading to receive the blessing of a name. Through Wales from North to South they travelled, and eastwards into the Welsh Borderlands, where two languages – English and Welsh – still shuffle back and forth from hill to hill, farm to farm. There, by Caer Caradoc, were magnificently exposed as rich a selection of fossils as could be wished for, and these had escaped the great forces that had so squeezed poor *Angelina*.

To begin at the beginning. The base of the Cambrian is one of the great watersheds in the history of life. This was when a host of animals we can recognise appeared in the fossil record, and when many organisms acquired shells and skeletons which could make common fossils. This proliferation in the rocks has been described as an 'explosion' in evolution, although I would prefer 'flowering' as a metaphor. Much more is known about Precambrian life now than was the case when the Cambrian was named: then, rocks earlier than Cambrian were thought to be virtually devoid of any evidence of life. Nowadays, there is much evidence of life in more ancient rocks – but it is not Cambrian life, not life with shells, nor obviously connected with life in our present seas. Our life – or the root of our life – flowered at the base of the Cambrian. I wish that there were some spectacular place in Wales where one could go to see clearly this origin of all origins, but there is no such place. There is an old quarry at Comley (Shropshire), not far from Church Stretton, which has yielded up some famous material of early trilobites and other fossils from a few feet of strata exposed there, but last time I went there it was overgrown and sad; it should be one of the holy places of British geology. Somebody had dumped a mattress.

A better place to get a feel for the early days of the Cambrian is the Harlech Dome. Harlech lies near the coast only a few miles south of Y Garth and its *Angelina*. To the east the ground rises steeply into wild country: into the Dome, so-called because the strata dip away from its centre in all directions. Slice the top off an onion and you will see the deepest layers in the centre surrounded concentrically by rings of the outer layers; this is a plan of the structure of the Harlech Dome. The oldest rocks are in the middle, and these are tough grits. These beds weather into benches, or natural terraces, each with a small craggy scarp, and, because of the structure of the Dome, the terraces are arranged into natural amphitheatres constructed on a gargantuan scale. Tier after tier rise up the sides of Rhinog Fawr, the mountain at the heart of the Dome. It is difficult to walk across the grain of this country, but there are a few natural paths more or less following the strike of the beds, which produce ancient tracks like the celebrated Roman Steps, which are ancient even if not Roman. Moorland tops the Rhinogs,

Cambrian Wales: typical scenery of the Harlech Domes, from south side of Mawddach Estuary, SW of Dolgelly, Merioneth.

birch and scrubby oak cloak the valleys. It feels like ancient country, and so it is, but the fossils that prove it are very rare indeed in the early Cambrian rocks. Younger Cambrian rocks surround the grits in a great swathe of open moorland country running from Maentwrog to Merionydd. These are slates and flags more richly endowed with what the old geologists called 'organic remains'. Trilobites are the first among them.

This Cambrian country east of the Harlech Dome is an area of broad swelling hills and boggy streams: sheep country again. It is terrible country to get lost in. When mist comes down, as it commonly does, the horizon shrinks to the nearest bluff. Maps become confusing. Is this the stream on the map, or merely one of a thousand anonymous trickles? Is this a footpath or a sheep track? You soon understand how it is possible to go round and round in circles. Sheep loom up occasionally, but their advice is usually not worth taking.

There is another great belt of Cambrian rocks running along the northern edge of Snowdonia – the Nantlle Slate belt. Unlike the Harlech Dome, this belt is only a few miles wide, and follows the usual NE–SW Caledonian trend. Here are some of the greatest holes in the world. Slates have been quarried in vast quantity: the Penrhyn Quarry at Bethesda is a monstrous gouge where great cliffs show stepped galleries as quarrymen have dug deeper and deeper into the vertical slate. Llanberis is just as grand. The roofs of nineteenth-century Britain were born here.

Slates are fine-grained rocks caught and squeezed in the Caledonian vice until their every particle lies at right angles to the pressure. They split, cleanly, along these new planes. The split, or *cleavage*, has nothing to do with a bedding plane – the original sea floor surface. In fact, slates often split at right angles to the original bedding planes, and it is obviously a waste of time to try to find fossils lying on such surfaces. But I can recall at least one experienced geologist beating on slates, and cursing his lack of success in finding 'organic remains'. Sometimes, a green or purple stripe running across a slate will reveal where the bedding originally ran.

The Nantlle Cambrian slates are slate *par excellence*. The cleavage surfaces are perfectly flat, they can be split into thin and uniform sheets, shaped to a standard, graded for colour. Slate grey is, not unnaturally, the usual colour, but purplish or 'blue' or green slates can be supplied on demand. The colour depends on chemical variations in the original mud. Splitting and dressing slate is a skilled craft. A century ago hundreds of men were employed at it, but there are few who know how it should be done today, and most of those that are still working are at Penrhyn. A small, well judged tap serves to cleave the slate precisely. The National Museum of Wales has an excellent little museum at Llanberis explaining the history of the slate industry, complete with a skilled and working slatesman.

Cambrian slates have contributed a whole aspect to the architecture of Britain, and not just to the Welsh terraces which line the valleys, and which were built to house those miners who deserted the fields for industry. Around almost every city from London to York, Victorian developments were roofed in Welsh slate. Aerial photographs of London suburbs from Penge to Norwood to Tooting show neat grey rectangles for mile upon mile: Cambrian roofs. The standard villa design is: brick house, porch, bay window, slate roof. Whether this is regarded as boon or curse depends on how much you admire the design or regret its ubiquity; whatever, such mass manufacture spelled the end of local architecture. But nobody could dispute the efficiency of Cambrian slate: lighter than stone tiles; reliably waterproof; more durable than thatch; easily repaired; no wonder it won the day over local products! There is even a subtle colour variation through grey to blue-green or purple – slate roofs are not *quite* monochrome. An artificial 'slate' has now been invented that is absolutely uniform in colour, and

*The gaping quarry at Bethesda, North Wales, where Cambrian slates squeezed in the vice of the
Caledonides were mined to roof the nation in the nineteenth and early twentieth centuries.*

next to a real slate roof it looks lifeless and matt. Real slate lasts. It will remain
true for another century that Cambrian grey tones, originating from vast north
Welsh pits, will paint the roofs of much of Britain.

If slates do not show fossils, how is it known that the Llanberis slates
are Cambrian in age? On a high gallery in the vast Penrhyn quarry near the
stratigraphical top of the slates there are some green slaty beds, and just here
(and only here) the cleavage of the slates and the original bedding on the sea
floor happen to coincide. When the slate splits in this site, for once, the ancient
sea floor is revealed. And a trilobite, *Pseudatops viola*, survived to prove its age. It
is a fugitive from the maw of Iapetus, squashed perhaps, but a triumphant messenger from the past. The great discovery was made by two quarrymen, Robert
Edward Jones and Robert Lloyd on April 9th, 1887. How observant these two
working men must have been! At the moment of discovery an unknown stretch
of mysterious slates – slates which clothed the roofs of a nation – became installed
upon the great scale of time . . .

Top   *Smoo Cave, near Durness, Sutherland. A cave cut into the limestones of Durness, a legacy of when the North-west Highlands were part of a tropical sea. See the poetical description by Heddle (page 55).*
Below   *The Moine Thrust at Knockan Crag, where the story of the Highlands began to unravel. The metamorphic rocks lie above the sedimentary rocks, at the change in slope, and the recognition that they were forcibly thrust there was pivotal to the understanding of the history of the British Isles.*

Above   *The Solway Firth. Scarcely the dramatic expression on the ground one might expect for the most fundamental geological division of the British Isles.*   Facing page   *The Highland Boundary Fault. The sudden change in scenery North-west of Lake of Menteith, Perthshire, is the record of this deep slice through the Earth's crust. The red Devonian rocks clearly show where the former mountains were eroded.*

Plants that flourish on the poor and peaty soils of ancient Caledonia: the club moss (top) a survivor from the Carboniferous; the sundew, Drosera, (centre), a plant that entraps insects in its sticky leaves to supplement its minerals; Silene acaulis (below) a charming cushion formed at higher altitudes. All three are found in Arctic floras.

Facing page   Top   *Dob's Linn, Moffat, Southern Uplands, the classic rock section for Ordovician and Silurian graptolites which swarm in some of the black shales. Enclosed by bleak moorland at the former northern edge of the vanished Iapetus Ocean.* Below   *Cader Idris in the distance, a mountain largely composed of Ordovician volcanics, with Cambrian rocks occupying the middle ground.*

Above *Skiddaw from the air. The rocks now making up much of the Lake District lay at the southern side of the former Iapetus Ocean.* Facing page Top *Old Red Sandstone (Devonian); St Anne's Head, Pembroke, South Wales. The richly red rocks colour soil and stone buildings in areas around the ancient Caledonian mountains.* Bottom left *The Rhynie Chert (Devonian) preserves exquisite remains of early plants, showing individual plant cells (a section has been cut through the rock and is shown enlarged).* Bottom right *The delicate pattern left by the fossil Devonian coral* Thamnopora *from the limestones of South Devon has been widely used as an ornamental facing stone. (Approximately life size.)*

Above *Carboniferous volcanic rocks create the comparative wildness of Arthur's Seat (here seen from Blackford Hill) within the sedate City of Edinburgh.* Left *The bark of the fossil tree,* Lepidodendron, *one of the important contributors to coal. Faint traces of its characteristic pattern can often be discerned in house coal.* Below *Bloody cranesbill, a limestone lover, seems almost too exotic for the British flora.*

To examine Cambrian rocks by the living sea one must go to St David's, at the tip of western Wales, where I saw the giant trilobite *Paradoxides* in the cliffs. A patch of Cambrian tips the peninsula on the north side of St Bride's Bay. But here, what you see inland gives little clue to the complexity of the geology so gloriously revealed at the coast. The peninsula has been planed off by the sea so that the landscape promises little of the richness of the rocks. The exception is the igneous intrusion at St David's Head which defies the sea with vertiginous crags to the west and makes a boulder-strewn hill to the east. But on the cliffs everything is revealed. Never a fault that has not been picked out by probing waves. There are anticlines, where the strata have been upwardly arched, and synclines, where they have been downwardly bowed. The rocks now lie at any angle but horizontal. Softer rocks floor bays, harder rocks jut out as promontories. Children can terrify their parents by skittering out over thin rock bridges between land and sea stacks. There are red mudstones and green or purple sandstones and hard black shales. The coast path swings back and forth, or climbs, or plummets according to the fortune of the rocks. The purple or green sandstones, especially, have provided local building material, seen in splendid ruin in the Bishop's Palace.

There is something particularly romantic about ancient fossils from the Cambrian: they have survived so much and for so long. 500 million years or more have passed since dark muds were stirred by the limbs of the sole-sized trilobite *Paradoxides* as it scuttled over the sea floor. Now, once again the creature tastes salt spray, but in its rocky transmogrification it is insensible to the chemicals that would once have alerted it to the presence of food. In the sea nothing remains of its kind, but the sea itself splashes and heaves just as it did upon Cambrian shores.

Probably the most common fossil alongside *Paradoxides* is at the opposite end of the size range for trilobites: these are tiny agnostids, no more than a few millimetres long. They are blind trilobites. Their heads and tails are very alike and a few flexible thoracic segments separate them, and they were capable of rolling up into a tight ball. So strange are they that it is obvious how they got their name, and specialists are agnostic about them even now. Blows have almost been exchanged over the question whether agnostids are, or are not really trilobites. They have been thought of as Cambrian parasites, or, more plausibly, as Cambrian plankton. Whatever they were, and whatever they did, they swarmed in inestimable profusion in Cambrian seas. They provide

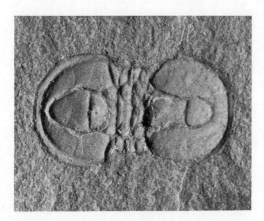

*An agnostid trilobite from the Cambrian —
agnostids were mysterious trilobites, unlike any
still living, as their name implies.*

me with an ideal opportunity to point out how necessary it is to have a hand
lens to be able to look at rocks and fossils properly; without a hand lens an
agnostid looks like a blob; under a lens it looks, well, like an agnostid. There are
hundreds of different kinds of agnostids. They changed through time, so just by
looking at these little enigmatica the expert can tell what part of the Cambrian
the fossils come from. This is practically useful in tracing correlative beds from
one part of the country to another, or even throughout the world. Some of the
same agnostids that lurk in the cliffs at St David's are found in Sweden, China,
Australia, even Siberia. I once had a visitor from Kazakhstan whose command
of English was rudimentary, but we got on perfectly well by merely exchanging
smiles and the names of agnostids.

Rising abruptly south-eastwards from the Cambrian slate belt, Ordovician
rocks make the true Welsh mountains, the greatest peaks outside Scottish
Caledonia: Snowdon and the Glyders, Cader Idris, Plynlimon. From the other
side of Wales, looking westwards from one of the vantage points in English
Shropshire like the Stiperstones, these mountains – the land of the Ordovices –
can be seen towering and remote, and then it is not difficult to imagine a tired
Roman centurion shrugging his shoulders, and putting off an invasion until next
year, or the year after. Like the Highlands, this part of Wales was deeply scoured
by ice only thousands of years ago. What remains now is a compromise between
ice and rock, blurred by weather and yet more weather in the millennia since
the permanent ice retreated. The legacy of ice is blatant in the scraped cwms
that scallop Snowdon itself, and in the sharp ridges between them. Here glaciers
scoured and gouged; dark hollows backed by steep cliffs remain, and small lakes
like Llyn Idwal nestle beneath, where once the ice lay thickest. Or in the Pass of

Llanberis the course of a vanished glacier, where the road now passes, is betrayed by all the rocky rubble dumped as it retreated. Six thousand years of weathering have not obscured these features. When there is a covering of snow in Snowdonia it is not so difficult to imagine vanished ice sheets even now.

The irony is that what Pleistocene ice has sculpted was itself the product of fire. Snowdon is largely constructed from Ordovician volcanic rocks. Tough and resistant to erosion, they survived naturally to take the high ground. Snowdonia is a vast, eroded syncline, and the youngest rocks at the centre now also lie on top of the highest peaks. Many of these Snowdon rocks were the product of volcanic ash falls which were sedimented beneath the sea. This was not inimical to life, which colonised – even flourished on – the volcanic sea floor in quiescent interludes. The very apex of Snowdon has one of these fossil beds, full of the shells of a ribbed brachiopod called *Dinorthis flabellum*, looking superficially like cockles (to which they are absolutely unrelated). This is the best example in the British Isles of the elevation of sea floor to mountain top. A purist would demand that you walk up to it, picking with your hammer as you go, but the Snowdon Mountain Railway makes things a lot easier for the rest of us. By a curious coincidence, the greatest

Llanberis Pass, in North Wales where Pleistocene ice has carved its way through Caledonian Wales, roads still follow.

elevated sea floor anywhere on the planet is also Ordovician: on the summit of Mount Everest there are Ordovician fossils really quite like those from North Wales. Our Welsh Caledonide example is hardly in the same league, but the processes that propelled *Dinorthis* to the top of Snowdon are the same processes that elevated Everest, driven by the ineluctable motor of plate tectonics.

The Glyders, Fach and Fawr: little and large in Welsh, have but a six metre difference between them. The top of Glyder Fach is stacked with jumbled slabs of volcanic rock where ice and weather have prised apart the solid rock until the slivers look like dominoes scattered by an irritable giant. The scenery here is as wild as anything in the Highlands. Cwm Idwal on the north face of Glyder Fawr was described by the naturalist Thomas Pennant (*Tour in Wales* 1781) as 'a fit place to inspire murderous thoughts' – thoughts which might have been brewed in the Devil's Kitchen (Twll Ddu). This is a dark chasm at the head of Llyn Idwal, developed along a joint plane in the volcanic rocks, down which tumble the waters from Llyn-y-cwn, a pool tucked among crags on the roof of Wales. The bones of the landscape project in jagged profusion from the carcasses of these mountains. Volcanic rocks are slow to weather. The walk from Tryfan to Glyder Fawr takes you along one of the most exposed ridges in Britain. There are more than 700 metres of Ordovician rock in Snowdonia, much of it lava and ash thrown from volcanic centres which on occasion erupted violently into the air, but at other times spilled more quietly beneath the sea.

The Ordovician was a time of volcanic tumult on the southern side of Iapetus, stretching from the rocks of Borrowdale in the Lake District, through Snowdon and Cader to the Berwyn Hills; and from Shropshire in the Welsh Borderlands and thence across country (with a few breaks) to Fishguard. The Welsh crust was strained, and stretched and faulted, and the volcanoes were the fiery response to the oceanic crust of Iapetus plunging down – and melting – beneath the continental rocks of Wales. On Strumble Head (Dyfed), near Fishguard, there are dark, basaltic rocks which were poured out upon the Ordovician sea floor. They make cliffs, thirty metres high, and have been sculpted by the sea into extraordinary, brooding forms. Look closer, and you will see that the lava flows are piled into rounded, billowing lobes, like some orgy of corpulent walrus or sea lions; these are what, less erotically, are termed 'pillow lavas'. Each 'pillow' shows how a fat finger of lava has solidified as it came into contact with the cooling water of the Ordovician sea. Between eruptions, there were even times when graptolites drifted past in the open sea, and now can be found fossil – looking like tuning

Above left *Great slabs of weathered Ordovician rocks on top of the Glyder Fach (From Pennant's* Tour in Wales, *1781) and* right *Pillow lavas: petrified, frozen billowing sacks of lava. This example is from the northern side of Iapetus, at Ballantrae, Argyll.*

forks – in little shale seams between flows; they give their unmistakable date stamp to lavas which might otherwise be confused with others from younger geological levels in the British Isles.

At the top of the Strumble volcanics at Garnwnder there is an ancient burial chamber marked on the map. A few miles inland, in the Prescelli Hills, there are many more reminders of the pre-Roman occupants of our islands. These open hills are the eastward continuation of the Strumble volcanics, and it was from here that one of the most emotive of all British rocks derived: the 'spotted doleerite'. This rock provides the 'blue stones' at Stonehenge. There is no reasonable doubt about this, because the rock is very distinctive, and not just because of its curious little spots, but even the most sophisticated modern analyses of its elemental composition agree. These huge blocks had therefore to move, one way or another, from Prescelli to the middle of the Cretaceous Salisbury Plain: massive loads over a long distance. There are those who claim that our ancestors were the agent; others claim that a Pleistocene ice sheet carried the stones on its back. The answer is of more than anecdotal interest. If it were a human achievement, it proves at a stroke that society then was organised, communicated over long distances, and could co-operate magnificently in common tasks of religious importance. This still seems to me the most likely explanation, for all the practical difficulties. There is meagre evidence for ice having carried blocks that large – or from the Welsh direction – into the Salisbury area. And surely it cannot be coincidence that western Wales is also rich in mounds, cairns and standing stones and all the arcane paraphernalia of these ancient peoples. I should add that the altar stone is a very different material – a sedimentary rock. A small piece of this rock yielded some minute fossils when it was dissolved in acid, which showed that it, too, was Ordovician, but we have not yet managed to 'match' it exactly.

The Menai Strait is a sea-filled slit between north Wales and Anglesey. Even a cursory glance will tell you that here is another major fault – it has the typical NE–SW Caledonian trend. The sea has found it out, and this is now a slippery place to try to collect fossils. Wales is crossed by several great faults; perhaps none of them has as profound an effect on the landscape as the Great Glen Fault, but each of them slices deep into the crust. Interesting things happen across the faults: Anglesey is a strange little island with a complicated geology all its own. It is not *quite* certain how it fits with the rest of Britain. Although there are Cambrian and Ordovician rocks there, they are different from those just the other side of the Menai Strait. The island was allegedly revered by druids, and perhaps they intuited its distinctiveness. In Shropshire, the valley that runs by Church Stretton marks another great fault. Everything acknowledges it: the railway runs along it, and so does the main road, and a succession of towns and villages are tucked into this corridor between the Longmynd and Caer Caradoc. The fault (actually, a complex of faults) can be traced onwards into Wales. Faults tend to sneak through valleys. They are planes of weakness like the faults in our own characters. Just as we can be sure that our own faults will find us out, so the elements will discover the faults in the geological landscape. They soon become hidden beneath alluvium and soil, and I suppose concealment is the natural tendency with human faults also. In the Ordovician, these faults in the Welsh Basin may have been like the faults that threaten California today – there would have been sliding and grinding movement along them, accompanied by massive earthquakes. The San Andreas fault has jerked catastrophically for millions of years, and so it must have been across Wales, when all that could flee in terror were trilobites and nautiloids, and the sky was dark with the smoke of volcanoes.

The Longmynd and Church Stretton, Shropshire. Ancient, Precambrian rocks dominate the landscape above the charming town, which lies upon a great but sleeping fault line.

To the west of the Church Stretton Fault lies the brooding hill of Longmynd. A tiny road takes cars across it; the ground rises almost precipitously behind Church Stretton and, as you climb, within minutes you are transported to a different world. Straggling sheep scatter across barren moorland, and the bare bones of the rock formations stick out from hillsides, and show vertical bedding. You might have been magically transported to the Highlands of Scotland – the ground *feels* ancient in the same way. And so it is. The Longmynd is a plug of ancient Precambrian crust dredged upwards along faults. Its Welsh side is also defined by another great fault. Its name perfectly embodies its in-betweenness, because 'long' is English, and 'mynd' derives from the Welsh 'mynydd', hill. The Longmynd is an *inlier*, an atavistic island emerging in a sea of younger rock. It is a tangible message from a deeper level than our present landscape. If faults supply a metaphor for our human deficiencies, inliers might stand for the collective unconscious. Deeply buried beneath the fossil-bearing rocks of England and Wales there is a foundation of older rocks from mysterious times: a more deeply hidden history, yet one upon which everything else is built. These Precambrian rocks were the islands on which Cambrian seas lapped near Comley. To judge from their steeply dipping beds it is clear that they have been folded and eroded

93

before the seas of Iapetus washed over them. From the top of the Longmynd you can guess this ancient history, as it might be an archetype that drifts upwards from some primeval level in the brain. Westwards lies the slouched shape of Corndon Hill, the igneous work of the Ordovician; eastwards lies the Ordovician of Caer Caradoc with Silurian Wenlock Edge beyond, and thence to younger England. The Cardingmill Valley is immediately below you, where a stream has sought to carve a sharp gash into the side of this long giant, asleep since the dawn of time. The mill used water power to comb sheep's wool, doubtless harvested from the animals that roamed the moors above, so joining in a strange marriage precocious industry at the beginning of the Industrial Revolution with the oldest rocks in England.

The Longmynd is just one of several atavistic islands, each crowned with heather moor and bracken, and supporting little but sheep and dreams. They were old when the land of the Ordovices was young. To the south, the Malvern Hills and the Wrekin poke up through the cover; to the east, in the very belly of England, Charnwood Forest (Leicestershire) is passed by the main northward motorway. These inliers give further hints of a hidden Precambrian landscape still older than our own. Water still flows from it. Malvern Water is one of the few mineral waters which has been bottled in Britain from before the Perrier invasion and there are springs at Church Stretton likewise related to faults and changes in permeability at the edge of the inlier. Whether these waters have real virtues I cannot say, although their analyses tend to suggest that they have a purity which stems from a lack of ingredients rather than an excess. Homoeopaths tell us that a few molecules, tellingly placed, can do wonders for the system. What I do know is that Malvern Water complements whisky very satisfactorily, and that thereby the English Precambrian and the Scottish Precambrian are conjoined in spirit.

The Charnwood Forest inlier should be mentioned, as a demonstration that the impossible happens. In 1957, a schoolboy, Roger Mason, saw the impression of a 21 centimetre long frond-like fossil on a steeply dipping bedding plane in an old quarry near the golf course. These, of course, were Precambrian rocks, and at that time the idea of large fossils being present in rocks older than Cambrian was decidedly novel. Were not all animals soft-bodied in that era, and therefore left no impression? Fortunately, Trevor Ford, at Leicester University, believed Roger Mason, and found more examples of the fossil. He named it *Charnia masoni* in a paper published by the Yorkshire Geological Society the following year. Once so named, *Charnia* became, as it were, visible elsewhere. Nowadays, there is a catalogue of places all around the world where *Charnia* has been found, and it is known to be only one of many large organisms which have left fossil remains in the later Precambrian, and which are particularly well-preserved in the Pound Quartzite in the Ediacara Hills of South Australia. Soft-bodied animals *do* leave casts if the conditions of preservation are right. Nobody was quite certain what *Charnia* was when it was discovered, and after more than thirty years experts are still arguing, although everybody is agreed that it is an organism. It is an intriguing thought that if Roger Mason had known more he might have seen less, might even have dismissed *Charnia* as an illusion. Maybe in this case the knowing eye might have been more inclined to turn a blind one.

I have collected beautiful trilobite fossils in sight of the Wrekin, a distinctive inlier composed of volcanic rock tough enough to make a steep-sided feature which is nearly conical when seen from the south. It still maintains a modicum of wildness, too, which is an achievement in the face of encroachment by the estates of Telford. There is a Shropshire expression, meaning something like 'cut the red tape': 'Ah,' they will say, looking conspiratorial, 'we don't want to go around the Wrekin'; and it must have been a considerable detour by horse and cart in former times.

A great stretch of Ordovician and Silurian rocks borders Cardigan Bay, surrounding the flanks and foothills of glaciated mountains: this is an impenetrable tract of bare hillsides, except where they have recently been planted with conifers, with few tracks. Broadly, the Silurian is centred on the upper Teifi valley and Aberystwyth, and has Ordovician to the north and south of it. The rocks are interminably folded and faulted, and if the steep, but rounded hills are reminiscent of the Southern Uplands, there is good reason for it, because the mid-Welsh hills are composed of similar, endless shales and grits. This is the centre of the Welsh basin, a great sink for all the clays and muds and grits on the southern margin of Iapetus. Like the Uplands on the northern side, this stretch of Wales is graptolite land, because their floating colonies drifted into the basin and were entombed in its muds. Thank goodness they were, for the graptolite zones have been the key to unscrambling the mysteries of the country.

I have an unreasonable prejudice against the Aberystwyth Grits, which are Silurian rocks exposed around the pleasant little town which gives the formation its name, lying close to the heart of geologically ancient Wales. This is because I spent hours in my youth whacking at the shales and not finding graptolites. The monotonous alternation of grits and shales looks like a club sandwich stacked to infinity. The grits (or greywackes) are the deposits of slurries of sediment which poured down into the Welsh basin; underwater slurries which travelled fast and far as a kind of turbid avalanche. Graptolites are preserved in the shales between, when the sea floor returned once more to quietude. If you want to see Silurian graptolites in abundance you have to travel a few miles inland to the gorge of the Afon Rheidol, where the sloping ledges of shales which crop out by the bouncing stream are full of wonderfully delicate specimens preserved in golden iron pyrites. There are graptolites that look like hair springs, or ratchets, or delicate brushes, or tiny hacksaw blades. Their shapes tick off time, more reliably than the coins of the Caesars.

The buildings of rural Caledonian Wales tend to utilise local materials, of which there are plenty for making rough walls: grits, volcanics, limestones, can all be turned to account. Geologists are grateful for the multitude of small pits that have been opened to exploit local stones, many of which have long since fallen into the hands of the Society for Dumping Mattresses in Old Quarries. The plan of the four-square farmhouse is as solid and sensible as stout Welsh shoes, with a front door in the middle that nobody ever uses (except perhaps the minister when there is bad news), and a door at the side that everybody uses, and a front parlour whose use is kept for best. The yard was formerly cobbled using small pieces of the same local stone, from which the farm outbuildings were also constructed, to make up three sides of a square. Whitewash is often applied to give the whole a clean and comfortable look. Formerly, a whole variety of local 'slates', and sometimes volcanic ashes rather than true slates, were used for roofing, and these included some very attractive pale greens and greys in large slabs; they very often survive on outbuildings but Cambrian slates from the Nantlle belt are the rule for re-roofing. In the farmyard there is always an amiable black and white sheepdog that chases car tyres.

It is clear that much of rural Wales was once comparatively heavily populated. The drift to the town was already noticed by George Borrow when he made his eccentric perambulations which he described in *Wild Wales* (1862), and it still continues today. Dry stone walls, which use such stones as come to hand, are often rather crudely made in comparison with their English counterparts, and now wander somewhat surprisingly across steep slopes in North Wales, where no-one has much of an interest in maintaining them. You can encounter similar walls in patches of secondary woodland, where they often lead to tumbledown cottages, often little more than a heap of jumbled Ordovician. A few of these cottages have been converted to weekend places for the town dweller, and it does seem a little perverse for these to become fire targets for nationalist extremists, when they would perhaps otherwise have decayed. There can, however, be a certain splendour in decay. On the Dyfed coast north of St David's artists like Graham Sutherland have been attracted to the ruins of the old slate workings there: strange, disembodied walls made from piled 'slates'. The cliffs here are one of the few places in Wales where the crow with the red bill, the chough, still thrives.

It would be disingenuous to deny that there is a good deal of rain in Wales. Because the air is also comparatively free from pollution lichens grow profusely. Old branches of deciduous trees in mossy hollows are perhaps the best places to find them, but all manner of stones are painted yellow or mottled grey or green with lichens. There are different species according to site and the different surfaces. It is as if the paint selected its own canvas. 'Glory be to God for dappled things', wrote Gerard Manley Hopkins; in the Caledonian landscape lichens are the mottlers and dapplers.

On the same stone walls as lichens there are ferns, like the delicate little maidenhair spleenwort (*Asplenium trichomanes*) and the polypody. Pennywort (*Cotyledon*) is the most typical flowering plant in the same situation; its name derives from the pre-decimal penny, which the leaves resemble in size and shape; I suppose we should re-christen it two-pennywort. In any case, this little plant points up the value of the Latin name as a stable point of reference in a mutable world. Its white flower spikes in June or July resemble miniature foxgloves; the real foxglove (*Digitalis*) is particularly profuse in hedges and at the bases of Ordovician and Silurian walls in Wales and Shropshire. There can be no more pleasing sight than a roadside verge in the late spring in the country where the Caledonides once towered, but where lanes now wander or buck up and down between stone walls: foxgloves, red campion, stitchwort, parsley, in a slightly undisciplined profusion which chemicals have not yet destroyed.

The mountain flora includes many things in common with the Highlands, and in this respect at least the former presence of Iapetus is not acknowledged on the ground. Here, too, are club mosses (*Lycopodium*) that would not have seemed out of place in the coal swamps of Carboniferous age. Above pastures (with the grass *Agrostis*) and moors there are Arctic relics, which grow at progressively lower altitudes the further north one goes. The purple saxifrage (*Saxifraga oppositifolia*) grows near the summit at Snowdon where it forms exquisite mats carrying profuse pink, five-petalled beakers of blossom. I have trodden on the same species at sea level on the Arctic island of Spitsbergen, at 80 degrees north. Incidentally, on Snowdon it often grows in crevices in rocks and thus could qualify for its Latin tag meaning 'rock breaker', but had it been found first in Spitsbergen it might have been called instead 'ground crawler' or some such, which goes to show that botanical names, while unchanging, may not always be etymologically apposite.

The geology of Snowdon itself produces a rather special flora, because the Ordovician volcanic ashes there are lime-rich: it will be recalled that the calcareous brachiopod fossils are also found there. In the steep cliffs above Llyn Idwal the flora is nothing less than luxuriant, so much so that a hollow near the Devil's Kitchen has been dubbed the Hanging Gardens. There is a mass of ferns and flowers, illuminated by the yellow Welsh poppy (*Meconopsis cambrica*), our own native representative of the genus that blooms in glorious blue on Arran. Some of the fern species are rare – they were over-collected in the last century, when some aficionados had a *passion* for ferns that can only be equalled by orchid fanatics today. You can understand why, if you visit the filmy fern house at the Royal Botanic Gardens at Kew. En masse they suggest every arabesque and variation on the theme of leafiness, subtle as a painting by Corot, profuse as a jungle.

But the rarity which will be sought by the dedicated botanist, which is confined to Snowdon and the Glyders in the British Isles, is the Mountain Spiderwort (*Lloydia serotina*). This is a delightful little bulb, with thin, almost grass-like leaves, which makes it nearly impossible to find unless it is in flower (June); the flower itself is four-petalled, and waxy white, almost orchidaceous, for all its modesty. It is named after its discoverer, Edward Llwyd (anglicised as Lloyd), a Welsh naturalist who was in correspondence with many notable scientists at the end of the seventeenth, and beginning of the eighteenth century. He brings us, naturally enough, back to trilobites, because it was the same Llwyd who wrote to Martin Lister in 1699 of the discovery of 'divers flatfish' in the rocks around Llandilo, South Wales, as reported in the *Philosophical Transactions of the Royal Society*. These 'flatfish' refer to a well known trilobite (*Ogygiocarella debuchii*) which abounds in the rocks exposed around Dynefor Castle, a minor stately home.

So Llwyd made the first observations on trilobites, just as he can be credited with the discovery of the Snowdon lily, albeit with less biological nous in the former case. *Ogygiocarella* has, in truth, an extremely passing resemblance to a plaice, although it did grow to be nearly as large. Llwyd also has a trilobite named after him: *Lloydolithus*, from the Llandilo (Ordovician) rocks. It was a neat little animal not much bigger than your thumbnail, with a fringe around it, and should Wales ever demand a geological emblem, would serve the purpose most satisfactorily. Both these animals escaped the distortion meted out to *Angelina*.

When I have been working on the Ordovician rocks in Wales, or in Ireland, passersby almost invariably enquire: 'Are you looking for gold?' It seems to be a common perception that gold is the one thing worth pursuing in geology. But, paradoxically, it *is* entirely pertinent to ask about gold in both Wales and Ireland, or anywhere in Caledonia along the vanished ocean of Iapetus. There *is* gold in the hills. There is silver, too, and lead, and zinc, and copper – all have been mined at one time or another, particularly in the eighteenth and early nineteenth centuries. Most of these minerals are found in veins, insinuated through the rock, along faults or joints. Gold mines are not busy now in Wales, but there is still gold to be had near Dollgellau; and in 1853 there was a Gold Rush within our own shores. It is doubtful whether many Klondike fortunes were made, but there was enough gold to keep one mine, Gwynfynydd, open until 1938. The gold for royal wedding rings still comes from North Wales. The best chance of finding some today is in the stream bed of the River Mawddach, where alluvial gold can still be panned here and there. This is gold which has been removed from its rock host and concentrated by erosion, over centuries. Nothing in Nature attacks gold, so it endures while all else wears away, which is the root of its symbolic incorruptibility. Its snootiness means that it is also found 'native' – as the pure metal, rather than combined with other elements which require smelting to remove them. Find gold, and you have instant fortune, without having to do anything; good as gold. No wonder it is the stuff of fantasy and plunder.

Nor is it a coincidence that adits run into the hillsides in just this part of the country. Gold lurked in the fumes from the end of Iapetus, its last exhalations. It is a very rare element in the crust of the Earth, concealed and diluted. It needs to be concentrated first and then deposited where it can be reached – a process of the most exquisite balance. Gold is deposited in veins which may follow faults or other weaknesses. Veins cut the country rock (proving that they have to be younger in age) which can be Cambrian to Silurian, according to the mine. The commonest mineral in the veins is usually vulgar quartz, white, opaque and very hard, and this will likely be the first thing you will find if you go gold scouting.

We must imagine the last grinding together of the two sides of Iapetus, and hot fluids carrying their precious gifts – distilled and concentrated – squeezed into the merest hair cracks in the country. The wonderful alchemy of the Earth resides in this concentration, as extraordinary in its way as anything predicated on the Philosopher's Stone, by which medieval alchemists sought transformation of base metal into gold. As a metaphor for the perfectability of Mankind the concentration of gold from the common crust is almost as potent.

The 'base' metals include lead: dull, heavy lead, seeming to have the weight of gold with none of its lustre. But it is valuable stuff, and in many places in Wales and the Welsh Borderlands it, too, has been mined. It often goes together with zinc, an association as close as Laurel and Hardy – not obligatory, but likely. One often comes across old workings with their spoil heaps, from which, on occasion, rare and valuable minerals are still being discovered, which were squeezed out alongside the more common ores. Workings which have produced many, many tonnes are in the village of Snailbeach in Shropshire, where an unreclaimed spoil heap bears testimony to an industry which lasted until well into this century. The surrounding hills include the remains of many old shafts, towards the village of Mytton, or at The Bog (now a study centre). Abandoned towers at the head of shafts are roughly built of local Stiperstones sandstone, and look as ancient as cromlechs, for all their recent origin: Whitegrit Mine looks as if it might have been dedicated to the worship of a pagan god. Mining has always been the toughest of ways to earn a living, and these Shropshire mines were exploited or abandoned as the market dictated. Greed and ruin drive hard bargains, and both drive mining. The crumbling towers are their monuments.

Between the Ordovician and Silurian there was a major extinction of marine animals, and this is connected with an Ice Age centred on the African continent, which produced an almost worldwide cooling. In Morocco you can see ice scratches just like those on glaciated rocks in Wales or the Highlands, except they are 440 million years older. The indications for this trauma in our own rocks are more subtle, although the change in the fossil animals is obvious enough. One might say that the end of the Ordovician glaciation marked the end of that far distant, most ancient world, where animals without backbones were overwhelmingly numerous in the sea, and the land was bare. Before the end of the Silurian, that step on to land had been made.

While Silurian mudstones gathered and grits plummeted into the basin in the midst of Wales, to the east there were tropical reefs with corals, and algae,

and all manner of wonderful life. Nowadays these reefs still make a feature on land as they must have made in the seas in the middle of the Silurian. This is Wenlock Edge. The Silurian limestones are much used locally for building, which gives Much Wenlock and its abbey ruins a 'Cotswoldy' flavour.

The poems of recollection written by A.E. Housman embrace something of the geological character of Shropshire; his 'blue, remembered hills' may not be the Edge itself, but I have stood on top of the Edge looking westwards towards Wales, and seen precise tones of indigo. Close to Much Wenlock there is an observation point on the Edge where it is easy to see corals and other kinds of fossils which built up the Silurian reefs in the limestones exposed there: white today – rendered and bleached by time into calcite – these reefs may well have been as colourful 425 million years ago as any living today. It is not too difficult to imagine the sea flooding in over the undulating country below until it washes against the cliff on which passing motorists stand today. Imagine the wash of breakers over the corals and algae; there would be trilobites in crevices just as there are crabs today, and there would be brachiopods and clams, on some of which there would be sea mats (bryozoans) growing. No matter that a zoologist would soon recognise that the occupants of these habitats were not the same as those living today: different kinds of corals, and bryozoans and so on. It would still be indubitably a reef. Ecology is more enduring than any particular organism; the actors change, but the show goes on.

The Wenlock reefs are the first good ones in the British geological record; there are more to come. The Wenlock rocks come to the surface again in an inlier at Dudley, Worcestershire, on a little hill known as the Wren's Nest. It comes as a surprise, because you approach the hill through the midst of urban streets: and there you are, suddenly, back in the Silurian. The sloping bedding planes of pale grey limestone give you an immediate view of the sea floor at that time. The fossils are starting to weather out; peeping out from the veil of hundreds of millions of years: corals, brachiopods, trilobites. The Dudley limestones have yielded hundreds of absolutely exquisite fossils in the past. They seem to weather out of the rock almost burnished. The trilobite *Calymene* was numerous enough to have acquired the common name of the 'Dudley locust' or 'bug', which shows that whoever christened it correctly guessed that here was an ancient arthropod. Nowadays, you are not permitted to collect, because the Wren's Nest at Dudley is a protected site, but you can still crawl about looking at a sea floor which thronged with life more than 400 million years before. You may come across the stalked eye of a trilobite, regarding you now just as once it spied upon the shallow Silurian sea, on the look-out for prey or foe.

The Lake District is arguably the most loved part of the Caledonian range. Close by the Solway Firth, it lay at the edge of the southern side of the vanished ocean Iapetus. Lakeland seems to break through the limestones of the Carboniferous which largely encircle it like the back of some monstrous giant hunched up in his primeval sleep.

The imprint of the glacier ice that vanished only seven thousand years ago is everywhere: in the deepened lakes, the 'waters' – Ullswater, Wastwater, Crummock Water – as obviously drowned glaciated valleys as any of the Scottish lochs; in the scalloped 'crags' at the heads of valleys as evidently the seat of ice as any cwm on Snowdon. Ice has left its shattered debris behind, it has carved the steep slopes, and the memory of its passage has instructed the roads on where they shall run. The valleys have gathered enough soil to support woodlands of birch and oak, which cannot subsist on the fellsides.

The marriage of ice and geology has made the scenery; Samuel Taylor Coleridge and his friends admitted it into the canon of what was aesthetically most desirable in the countryside, and so it has remained. The rocks are of early Ordovician age in the northern fells around Skiddaw and Keswick. They get younger southwards, so that great Helvellyn, Scafell and the Old Man of Coniston are made from late Ordovician volcanic rocks. This is the story of Snowdon again: a great volcano sometimes exploding like Stromboli, hurling out fiery, ashy clouds incandescent with heat. Pile upon pile were thrown out – five hundred metres altogether – and the Ordovician sky must have been black with smoke and the air foul from the exhalations of fumaroles. Today, the lavas form the crags, while the benches between record where the ashes fell into the sea. Appropriately, this is where the wildest country lies today, where those who have thought of the Cumbrian mountains as mere hills have frequently had cause to thank the Mountain Rescue Service. This volcanic area was the centre for the glaciers of the ice age, too, and a glance at the map shows how the glacially deepened lakes and meres radiate outwards from the ancient eruptive centre, almost as if they had been the traces of lava flows. In this fashion the memory of a volcano vanished for 450 million years lingers on in hidden ways. In the west, around Eskdale, a great Caledonian granite welled up from the depths of the earth, baking and altering the rocks it abutted. Southwards again, by Coniston Water and Lake Windermere, into

The Wren's Nest, Dudley, Worcestershire, where Silurian rocks on the slopes show glimpses of rich and varied marine life, including corals.

Furness and the Howgill Fells, and there are Silurian flags and shales like those around Aberystwyth. Graptolites have written the narrative of time in many of the sedimentary rocks, as they did in Wales, but above the volcanic rocks there are a few places for the trilobite lover also. For another kind of rockcomber there are old mine workings which still yield minerals from their spoil heaps, perhaps an even greater variety than in Wales; copper and tungsten were mined here and maybe one day will be again. Through all this richness of Lakeland rock and history Mr Wainwright has charted a matching variety of walks, and I will not inadequately attempt to dog his footsteps.

My own experiences of trying to find something in the shales around Skiddaw have not been very rewarding. Mile upon mile of Ordovician 'slates' are there, but all trace of past life seems to bury itself impossibly deep in the rocks as soon as it senses my approach. Great slithery piles of scree yield nothing. It is just that things are very rare in the 'slates' of Skiddaw. It may be a marginal consolation that they are even rarer in what are supposed to be somewhat equivalent rocks on the Isle of Man (Manx Slates), where nothing more than a solitary sponge has turned up for all the years of searching. Perhaps it is better to enjoy what Coleridge enjoyed and forget these particular rocks.

Plenty of tough local stone has resulted in attractive farmhouses throughout the Lake District, which are tucked into valleys between crags. The four square, double-fronted house is often whitewashed, while attached byres and barns are left to display their geology. In particular, there are beautiful green 'slates' from the Ordovician volcanic sequence which are used to fashion some of the most beautiful roofs in England. They complement the richer green of the fields. These slates cannot be cleaved very thin: as a consequence such roofs are also probably the heaviest anywhere, and require massive support from beams and lintels. The grading of the 'slates' from small at the ridge to large at the eaves is like the Jurassic roofs of the Midlands; moss and lichens add further subtle tones. Barns are often constructed from slabby rocks piled and mortared cleverly together, and many show the curious feature of projecting slabs, which appear to make a kind of staircase on the outside of the building. Stone walls complete the picture, which may be made of slates, or from stones picked from the fields – tough lumps of granite perhaps.

There are wonderful names covering this part of the hidden landscape. These are not merely becks and gills and thwaites. One thinks of Dollywaggon Pike, Crinkle Crags, Swineside, Ling Thrang Crags or Catstycam. I have tried to find fossils (and failed) on a hill called Trusmadoor. Then there is Great Cockup.

This igneous intrusion provides a prompt for mentioning Sod's Law, which is more pervasive in geological life than elsewhere. In the Lake District, its existence is proved; whenever you drop a chisel it falls with a plop into the most inaccessible part of a rushing beck. Or when at last you find some remain of ancient life in the ungrateful shales it will break into a dozen fragments, six of which will bound down the fell to come to rest in a pile of scree, where they will merge forever within a horde of indistinguishable chips.

It is no indulgence to complete the story of the land of the Caledonides by returning to my favourite bit of it – Shropshire. The Ordovician volcanoes here are less elevated than in the Lake District or Snowdonia but I defy anyone to climb Corndon Hill without feeling the immanence of the remote past. Nearby, the Stiperstones ridge chases northwards across the country to prove that a geological map can be read on the ground as readily as a road map of the A409 if only you learn how to *see*. The tough Stiperstones Quartzite outlasts the rocks to either side; the white stacks on the hilltops are monuments to this endurance. Look closer. There are the tubes of some Ordovician worms that made their homes in these ancient sands, not unlike the Cambrian pipe rock at Durness. Heather and ling grow thickly on the hillsides: heather, which seeks out poor soils. Listen, and you will hear the whirring of the grouse, which feeds on heather, and therefore likewise follows the geology. On the flanks of the hill you will notice, picked out among the heather, tumbled blocks of white quartzite, ranged in lines. If you try to walk across the heather you will likely twist your ankle, for it is treacherous. These are stone stripes, a relic of the heave of the soil when the winter ice briefly melted, upon a ground deeply and profoundly frozen in the grip of the Ice Age; once again, the Pleistocene provides the seasoning upon the geological main course.

Not far away, the upper reaches of the River West Onny follow what is to my eye a nearly perfect valley. The Stiperstones Quartzite is on the skyline. The stream itself unerringly follows a fault separating a Precambrian inlier from Ordovician – because where there is weakness it shall be sought out. Not that this is to be regretted, because the valley is green and steep-sided, and the river has been dammed to form a small ornamental lake to set off a perfectly proportioned country house, Linley Hall, at the head of the valley. The woods that line the valley are full of bluebells, and such conifer planting as has been done is not oppressive, as broad leaved trees remain in patches, perhaps to preserve the habitat for pheasants. There must be many valleys like this one, shaped by geology, nurtured and used by men: various, unspoiled but not cosseted. To discover one like it is one of the joys of Britain. The value of such a place resides in memory deep as geology itself, a landscape that cannot be corrupted by time, nor diminished by the accidents of history.

The Stiperstones, Shropshire. Ordovician quartzites are hard enough to stand against the assault of weathering to remain as strange piles and monoliths.

Culzean Castle, Ayr. A castle where artifice has augmented nature to produce a romantic ideal. General Eisenhower was based here in the Second World War.

8

The Red and the Black

MOUNTAINS BEGIN to die as soon as they are born.

As soon as the Caledonian Earth's crust crumpled and thickened – the moment at which the mountain chain was given birth – then there was an onslaught of all the processes of erosion, seeking to reduce it once again. The higher the mountains grew, the worse the attack, because frost and ice are merciless shatterers and grinders of all rocks, no matter how hard the rocks might be. When the Iapetus ocean was consumed, the Caledonides arose magnificently to signal the fusion of the two halves of our islands. Then, what tectonics had elevated the elements immediately sought to diminish. But at this moment also, and for the next 400 million years, the narrative of our islands became a unified text. The Highlands were bound to the Lowlands, and the Lowlands to England and Wales. Our identity had been annealed.

This is how the Caledonian mountains wasted: great tumbling boulder-strewn rivers, gorged with flooding, tossing, rolling, breaking, rounding rocks, pummelling the weaker ones to nothing, seeking out micas to flake them away, assaulted rock ridges that resisted; all this under the glare of the sun, not halted by any stands of trees, for there were none. Debris cascaded into basins, some of which were bounded by faults, which moved to accommodate the mass of waste. The degradation of the Himalaya today provides a very rough picture of the process. But then, beyond the montane basins, were less tumultuous plains, built out in front of the mountain range as a skirt of its own waste. Here, violent rivers became gentler, and began to drop some of their finer sediments; pools of shimmering, fresh water amalgamated into lakes. And slowly the mountains could be shifted across the plain – grain by grain, year by year – towards their eternal resting-place in the sea, the sea from which they had themselves once grown in the last great cycle of the Earth.

The wasting of the Caledonides began in the Silurian in a few places, and was general in the Devonian. Now look to the map to see where these basins are: as you would expect, they lie adjacent to the Caledonides, in some places blanketing the ancient folded or metamorphosed mountains, in other places banked against them. The debris of the Caledonides is called the Old Red Sandstone.

These sandstones are often coloured a particular species of rich red. There is nothing angry about this colour, which often carries a hint of purple or carmine. The colour is passed on into the soils of Old Red Sandstone country, which are deeply ruddy. This is the colour of Herefordshire, lingering just beneath the lush green of meadows, revealed at every scraping or trench. It seems to carry further into the colour of the Hereford cattle that graze the rich grass, which are warmly red-brown mixed with cream, as if a transmutation of the elements could pass from soil, through grass, eventually to stain the hide and hair of the beasts of the field. The source of the colour is ordinary enough: iron.

Iron is one of the abundant elements of the Earth, mated with nickel in the core, combined with other elements in dozens of minerals in the crust: iron is a master of disguise. The red colour of the Old Red Sandstone is from the iron in its *ferric* state. This is what happens to the element when iron compounds weather in the presence of abundant oxygen: it is the rust* of the Earth. The darker, green, ferrous, form of iron is more common in marine rocks; the red rocks speak of erosion under the sun and in the air. The redness of the Old Red Sandstone is a true measure of the Devonian world; were it not for the abrasion of a mountain range which was already a distant memory when the first dinosaurs walked the land there would be no red soil through the dairy country of the Welsh Borderlands.

The 'Old Red' was named to distinguish it from the New Red Sandstone, which is of Permian to Triassic age: younger, brighter red rocks. 'Old Red' appeared on the first geological maps, so distinctive was its colour, and it is now divided into many different formations locally, which need not detain us. It has to be said that the 'Old Red' is not the easiest part of the geological column in which the historian of life searches to find ancient traces. It is no use attacking a random

---

*The chemistry of these processes can be quite complicated, with various oxides and hydrous oxides and carbonates of iron being the colouring agent, depending on the circumstances of weathering.

outcrop with a hammer expecting to find fish; disappointment will result. The fossils hide in special places, thin seams, a few quarries. For the most part, the 'Old Red' is intransigent stuff.

Each of the 'basins' in which the Caledonian waste accumulated was really a separate story: the rocks are different ages in detail. Comparable basins are present along the whole length of the Caledonian mountain chain, from Norway to the Appalachians.

The most northerly basin in Britain takes up its north-east corner: Caithness, John O' Groats, and the Orkney islands northwards across the Pentland Firth. Some of the rocks here show much less evidence of rich 'Old Redness' than other Devonian regions of the British Isles, so much so that the old name could seem a misnomer in places. But what a contrast these rocks of the Orcadian Basin make to the igneous and metamorphic rocks that comprise the endless moors and Beinns to the west. They are the first rocks in this book that are flat and unchanged. Almost everything thus far seems to have been squeezed and buckled – or at the very least tipped on its end – but the cliffs at Dunnet Head or Hoy are ribbed, and horizontally so, and each horizontal rib is the trace of a sandstone bed. We can look at a year of Devonian time without having mentally to unstretch, unbuckle or unbake. On top of Caledonian turmoil, this is the aftermath.

Not that it is an easy matter to reach some of the Devonian days, because the cliffs are often stupendous. There are joints at right angles to the bedding, and these define the cliffs: sheer, daunting, and loved by guillemots and kittiwakes and the foolhardy. The onslaught of the sea is without remission, and joint planes have been worked and eroded until sea stacks have eventually become isolated from the main outcrop, standing like eroded colossi against the elements. There is The Kist on the west shore of South Ronaldsay; on the island of Mainland, surely a confusing name, there is the Castle of Yescanaby; while on the mainland, the real mainland, there are stacks at Duncansby and Holborn Head. Most famous of all there is the Old Man of Hoy, which is to the rock climber what Tartini's Devil's Sonata is to the violinist – 137 m high, a petrified pole more than an island, improbably attenuated. This, too, is horizontally-bedded sandstone, the beds stacked up like a pile of children's bricks which teeter yet hold together. On top of the Old Man there is a patch of grass where seabirds can contemplate the peculiarities of a creature that would climb it the hard way. This patch is on a level with the plateau of the coast nearby; all over the Orkney islands the Old

Red cliffs are backed by a plateau, a plain rendered flat by the action of the sea, and now uplifted. Houses crouch on top of it, dwarfed alike by the elevation of the cliffs and the wide sweep of the horizon. The distinctive landscape continues northwards to the island of Westray – and the Devonian rocks run on still further beneath the sea.

Wildness has not prevented the habitation of Orkney since prehistoric times. There are standing stones in plenty, and there is The Ring of Brogar. Because the natural sandstone is so often horizontal, or nearly so, these vertical sandstone blocks have an especially deliberate ambience, as if to announce: 'Here, in this bleak place, we have made our mark.' Strangest of all is the Dwarfie Stone, a Neolithic burial chamber within a single, huge block of Hoy Sandstone. It seems as if a stone-eating maggot had chewed the block out from the inside. On the Caithness plain there are more antiquities, and the Old Red has made comparatively low ground which is more fertile than the Precambrian wilderness to the west and south. Cattle and crops have probably always done well here, and the many little roads and small farms – built square and sensible of local stone – are just the last in an occupation that stretches back to the time when stone circles were the centre of ritual. It is extraordinary how geological reality was recognised by men whose knowledge, or at least whose *scientific* knowledge, must have been next to nothing. But the language of stock or crops thriving or failing speaks louder than any petrological analysis.

If you look closely at the Orcadian sandstones, it will not be long before you discover beds with surfaces covered in fossil ripples. Maybe nearby there will be somewhat muddier beds covered with polygonal cracks – sun cracks, most likely, the legacy of one particular sunny day 400 million years before. The coarser sandstone beds will be full of finer flutes or scratchings which run obliquely to the horizontal bedding planes. This is *cross-bedding*, which is the legacy that migrating sands, moving as waves or ripples beneath water or wind, leave in the rocks. All these signs tell us clearly that the Old Red Sandstone accumulated in shallow water – exposed to the sun at times – rivers or floods, perhaps, with no hint of the sea. The Caledonian waste was fed into fresh water 400 million years previously, and the products of that ancient erosion now feed the Aberdeen Angus.

The freshwater habitats of the Devonian were frequently unsuitable for pre-serving the remains of animals. But near Thurso there was a large freshwater lake, which was teeming with Devonian life. The Caithness Flags from Achanarras have yielded profuse remains of fossil fishes. There is nothing red about the 'Old Red' that contains the fish. The flagstones are dark grey, and break out in perfectly flat slabs: the fishes show up black on the surface, like a kipper skeleton spreadeagled on a plate, but tarred by the patina of age. Clearly, by this middle Devonian time, a variety of animals had moved from the sea to exploit the opportunities that were to be found in rivers and in lakes. These Scottish specimens were collected most assiduously by a local baker from Thurso called Robert Dick (1811–66), and were to become world famous through the writings of Hugh Miller, whose first book, *The Old Red Sandstone; or, New Walks in an Old Field* (1841) is one of the classics of geology. There is a *Dickosteus* and a *Millerosteus* (to say nothing of the 'sea scorpion' *Hughmilleria*) amongst these fossil fish to commemorate the achievement of these early scientists.

What a gallery of strange creatures they were! Many of the fishes were thickly armoured, especially on their heads and trunks. There was a considerable range of them, more than a dozen genera, proving that the fish must have had an earlier evolutionary history elsewhere. The most common larger fish is probably *Coccosteus* – an 'armoured shark' – with a jaw which could gape widely, with grasping or biting teeth, and which must surely have been a predator. Other species included a lungfish, whose relatives live on today in the swamps and rivers of Australia: these were air breathers. And there was a 'lobe finned' fish, distantly related to *Latimeria*, the coelacanth fish whose discovery alive and well in the sea was one of the exciting zoological events of this century. There were swarms of a small fish called *Palaeospondylus*, thus proving once again that the length of the name is inversely proportionate to the size of the animal. The Devonian lake must have thronged with life, and it seems extraordinary that some of this life can be related to creatures still alive today. From time to time the Devonian waters became poisoned. This may have been caused by a proliferation of the wrong kind of algae. I saw my goldfish die this way one hot summer. Or possibly the poison came from a volcanic source; there was still much volcanic activity in the Devonian. In any case, the lake that had so recently been the cradle of life became its coffin. Graveyards of fish fossils remain to tell the story.

Another great basin of Old Red Sandstone surrounds the Moray Firth. Inverness, at the end of the Great Glen, is built upon it. Surrounded by Caledonian granites and metamorphics, this area is like a fertile island in a sea of moorland. Southwards again, a great band of Old Red Sandstone takes up virtually all the rolling ground on the north side of the Midland Valley of Scotland, and immediately to the south of the Highland Boundary Fault. It is a dozen miles wide or more for most of its length, extending from Glasgow in the east to Perth and Dundee in the west. I have already described the eastern end of the 'Old Red' at Loch Lomond. This is where fans and screes and aprons of debris poured off the mighty Caledonian range to the north, banked up against the Highland Boundary Fault. All along the Fault hill after hill of conglomerate – rocks often as richly red as one could hope for – are monuments to the powers of erosion under sun and under torrent. At the other end of this belt of outcrop the Old Red Sandstone reaches the sea between the mouth of the Tay and northwards to Stonehaven. At Whiting Ness, near Arbroath – famed for its 'smokies' – there are conglomerates and sandstones exposed by the shore that were laid down in fast streams, and jumbled mudflows. There is even an unconformity *within* the Old Red Sandstone which is beautifully exposed here, and is proof that the Caledonian world continued to move and shift; an insecure world harbouring life that was itself changing, transitional, adventurous.

Away from the ancient mountain front, there were 'Old Red' lakes and rivers, more conducive for the preservation of the traces of past life. It was in the freshwater rocks now preserved in the old stone quarries of Carmyllie, near Arbroath, that the arthropods reached their apotheosis. Arthropods include all those animals with jointed legs, 'creepy crawlies': spiders, insects, millipedes, shrimps, lobsters, trilobites – the most diverse group of all living animals.

But arthropods have an upper limit on their size. J.B.S. Haldane, the one true master in the popularisation of science, pointed out long ago that there is an appropriate size for most organisms. For the arthropods, that size is small. This is because arthropods are encased in a skeleton which lies on the *outside*: there are limits to the weight of body that can be supported on tubular legs without them buckling; and because breathing takes place through the surface of the animal, simple geometry means that the larger an arthropod gets (its volume being in proportion to the *cube* of its length) the disproportionately larger the area necessary to ensure breathing, let alone activity. I am afraid that the movie featuring *The Three Stooges* on the planet Tharg where they were threatened by a tarantula the size of a small house is likely to be inadmissible. So *Pterygotus anglicus* from the 'Old Red' is the ultimate arthropod – at full stretch it was as long as a man. Admittedly, some of its length was supplied by stretched-out arms at the front, each equipped with nasty nippers. It is a kind of sea scorpion, or eurypterid, not an entirely appropriate common name since this one neither

The giant Devonian sea scorpion, Pterygotus, *must have ruled at the top of the food chain when the Caledonian mountains were at their highest. This specimen is nearly two metres long. Old Red Sandstone, Carmyllie, Forfar.*

lived in the sea, nor was a scorpion. But what a predator it must have been! Its vast size, at least for an arthropod, meant that it must have spent much of its life torpid, but watchful, waiting for something edible to turn up, and when it did, grasped it in a quick turn of speed. *Pterygotus* had a paddle-like tail, no doubt a great asset when it needed to turn up the knots. It seems unlikely that it could have maintained a dash for any length of time, because of the problems of size and oxygen. Some of its cat-sized relatives were closer to true scorpions, even to the extent of having a sting at the tip of the tail.

Further patches of Old Red Sandstone come to the surface through the Carboniferous as inliers within the Midland Valley. One centred on the small town of Lesmahagow is famous for its fossils ('Old Red' rocks start in the Silurian here) which include swarms of smaller sea scorpions, minnows to the pike of *Pterygotus*. There is a patch of Old Red Sandstone on Arran. Devonian forms the coast between Ayr and Girvan. This is a fine stretch of hill country, and halfway along it is Culzean Castle. Scottish castles nearly always have a dramatic setting. This started from necessity, because an elevated site, or one on a promontory, was the better defended. It must have continued by design, an awareness of what a castle ought to be as well as must be. Culzean started as a functional castle, and was then perfected by Robert Adam in the eighteenth century to be the definitive, romantic castle. Culzean has parapets that sit on top of a precipice to the sea. To peer over is a test for vertigo. Look northwards and the cliffs tumble to the Devonian rocky foreshore. But the landward side is almost demure, a made garden with rhododendrons and exotic plants that grow in the protection of the walls. It is a strange juxtaposition, encapsulating or symbolising Scottish history.

Old Red Sandstone makes another band across country north of the Southern Uplands Fault, thereby defining the southern edge of the Midland Valley, and providing something of a mirror image to its northern edge, although the geology is more complex and discontinuous. The Pentland Hills (Midlothian) are a delightful range of wild and open hills, often rounded, and covered in moor-grass, with valleys between that may be glacial drainage channels dating from the Ice Age. The valley floors support fields dotted with sheep, but the hills hardly support a tree. This is one of the best places in Scotland in which to find the remains of life of the Silurian and Devonian periods. Several of these fossils were discovered by Dr Archie Lamont.

To the south of the Pentland Hills there is a small town called Carlops, and Archie Lamont lived here for many years in Jess Cottage. It is sometimes claimed that scientist-eccentrics were nineteenth-century figures; the twentieth seems to impose a kind of technological boffinism accompanying worthy dullness. Archie Lamont gives the lie to this. He was magisterially, splendidly eccentric; a Scottish Nationalist who was expelled from the party for being too extreme. He wrote upon almost any geological topic, but particularly upon matters palaeontological. On everything, his views were heterodox. To promulgate his work he founded his own journal, the *Scottish Journal of Science*, which was, naturally, substantially filled with articles by Dr A. Lamont. His cottage in Carlops was a place of monumental squalor, and there he kept specimens – some of them unique – which he had collected from the Pentland Hills. Few were permitted to see them, and then only persons born on the Northern side of Iapetus. I met him only once, not long before his death in 1985, at the Royal Society of Edinburgh, where his tall, wraith-like figure wrapped in an antique coat was a little sad. He smelled vaguely of paraffin. He pressed some of his writings upon me. These included poems in Scottish dialect such as 'The Lallans Cats', and splendid polemics revealing a comprehensive range of scholarship as well as a lively pen. I remember 'William Blake and the stone of destiny', a title as grandiose as any I can think of, but then, he described some trilobites in a paper entitled 'Scottish dragons', which shows a facility for a compelling title, if not verisimilitude. Even if he was a bit batty, particularly in his later years, who can pass through the Pentland Hills without recalling him?

Through much of the Midland Valley the Old Red Sandstone sediments are accompanied by volcanic rocks, proving that the Earth was far from settled after the Caledonian upheaval. The volcanic flows make major features north and south of the Firth of Tay: the Braes north-west of Perth, and the wild Ochil Hills to the south, just as volcanics are responsible for the rockier crags in the Pentland Hills. Along the Tay Valley floes of resistant andesite often form bluffs, craggy and speckled with trees on the scarp slope, where the dip slopes weather to form good agricultural land.

Volcanoes must have been another hazard for creatures crawling and swimming in the Devonian lakes and streams. They may have provided bonuses, too, because volcanoes often unlock nutrients hidden deep in the Earth. At one site, now in a little outlier of 'Old Red' near Rhynie, south of Strath Bogie, Aberdeenshire, a hot volcanic spring bubbled up over the ground from time to time, its waters charged with silica, which became deposited as a hard crust. But this spring was a wonder. What it bubbled over was the earliest peat, and what the silica preserved were land peats, petrified, cell-by-cell. The Rhynie Chert, dug from an unexceptional grassy hillside, preserves what is seen hardly anywhere else, a time of adventure, when plants began to clothe the land. They are simple plants without true leaves – more shoot than shrub. But they could creep over the land, create little spreads. The greening of the land was under way. And once a green canopy was there other things could follow . . . and so they did.

There are mites in the Rhynie, the tiniest arthropods – the inverse of *Pterygotus*, though it lived not far away – and there is an equally minute insect, a springtail. If you shake the rotten vegetation around a peaty pool, in the same hillside where the Rhynie Chert crops out today, the chances are that some little, dark, living specks will hop out on to the water: springtails. Imagine! Four hundred million years of continuity . . . These tiny companions of primeval plants survived to spread around the world, and back again, to the very spot where they had started.

Springtails similar to these living ones can be found in the Rhynie Chert (Devonian), an example of the inconspicuous animal enduring while more spectacular ones perish.

The Old Red Sandstone buries the Ordovician and Silurian folds of the Southern Uplands along their eastern edge, through a beautiful stretch of the Scottish Lowlands running southwards across Berwick and Roxburgh to the Border Country. Near the northern end of this outcrop I have mentioned previously Hutton's famous unconformity at Siccar Point, where horizontal Devonian ignores the agony of the folded Silurian beneath, irrefutable proof of the wasting of the Caledonides. The Old Red provides some of the gentler ground running north of the Cheviot Hills, which is where volcanics of the Devonian once more make the high ground, until they are themselves over-topped by the implacable, intrusive granite of The Cheviot himself. Given its comparative accessibility, it is perhaps no wonder that the sandstone area was one of the main sites for skirmishes in the Border wars. Jedburgh Abbey was destroyed four times by the English since its foundation by David I of Scotland near the time of the Norman Conquest. Now, it remains in ruins. At Jedburgh, too, Mary Queen of Scots lived for a while in 1566, and Queen Mary's House remains, casually historic, near the centre of the town, today used as a museum. All these older buildings were constructed of Old Red Sandstone, solid slabs and blocks jigsawed together to achieve a solidity that sought to defy the humour of the times. The blocks weather patchily, and thus contribute to a mottled surface running through all the subtle colour tones 'Old Red' can offer. Near the centre of the outcrop the English Borderers fought the Douglases in the bloody Battle of Ancrum Moor (1545). There is alleged to be a cluster of caves by the old bridge spanning the stream at the edge of the village hollowed out from the sandstone itself, which is unusually soft in this region. The inhabitants, children perhaps, shuddered within, waiting for the next waves of violence to pass. It is one of the more grisly claims to fame of the Old Red Sandstone to have provided the theatre for so much bloodshed, while offering all too inadequate protection from human folly.

This historic landscape is Sir Walter Scott's fiefdom, and his beloved house of Abbotsford is situated on the Silurian rocks a mile or two westwards of the unconformity with the Old Red Sandstone. It is a most curious confection, part gothic, a little bit Elizabethan, part sensible Victorian country manse. Scott was another of those prodigiously energetic nineteenth-century Scotsmen, a match for Sir Archibald Geikie, the director of the Geological Survey, whose level of output induces admiration and depression at the same time. His drive made him pursue and collect every *quaich* (an antique whisky nip with little handles) north of the border, which now fill curiosity cabinets at Abbotsford. I have the same mixed feelings about the border castles that Scott admired so much: Hermitage Castle (South of Hawick) or Norham (Northumbria). We admire the setting, snuffling about the past, treating them as sepulchres almost, until we are caught out by some detail that reminds us that they were built on the bones of oppression and consecrated in blood.

The Welsh Caledonian mountains cast their Old Red Sandstone debris eastwards and southwards, where it comprises the widest belt of these rocks in the country. From Ross-on-Wye west to Llandilo is sixty miles, from Bridgnorth (Shropshire) south to Newport, seventy, and all of that on 'Old Red'. Its outcrop passes to either side of the South Wales coalfield, which, being a great syncline, assures us that the Old Red must scoop beneath it, as a hand might hold a bowl. North of the coalfield, it takes the high ground to make up the Black Mountains and the Brecon Beacons, including a National Park, and thence north in its more domestic mood to underlie the body of Herefordshire.

At the northern end of its outcrop, in Shropshire, the change from marine conditions to brackish or freshwater deposition happened within the Silurian, as it did in parts of the Midland Valley. It would be impossible to exaggerate the importance of the Silurian-to-Devonian 'Old Red' rocks surrounding the Welsh Caledonides in our understanding of the history of life. Here land plants (*Cooksonia*) even older than those of the Rhynie Chert flourished next to ancient streams. The millipedes crawled away from water, and the oldest spider has recently been discovered. There were many different kinds of fish in fresh and brackish water, especially primitive, jawless ones, distantly related to the living lampreys. By the end of the Devonian there were approximations to forests, and there were ungainly, four-legged animals lumbering through undergrowth rich with life. The origins of much of this – the semblance of our own world – lie within our British rocks.

The step of our own ancestors from water to land is part of this story. An image of it is lodged in my memory, derived, I suppose, from a childhood picture book. A fish flaps awkwardly on to the mud using its fins as props; they change conveniently into stubby, five-toed legs – meanwhile, the fish has developed lungs, so that it can breathe the air we still breathe. And off it waddles to give rise to Man. Somewhere else in the bottom drawer of memory there is an impression that the ancestor to all this drama was a lobe-finned fish, not unlike the coelacanth fish *Latimeria* which still survives. Its stumpy fins seem somehow ready to transmute to legs. But all this may be so much twaddle. Fossils from the 'Old Red' of Greenland, in particular, have recently shown that early animals with legs had *more* than five toes, that legs, and very probably lungs, developed in advance of the march on to land. Other anatomists have claimed that it is the lungfishes – not the lobe-finned fishes – which are closest to all four-legged animals. The final story is still far from clear, but what is clear is that 'Old Red' rocks lay at the hub of it all . . . This is the very beginning of a new world*.

When I drive westwards from Monmouth and see the Black Mountains looming on the skyline near Abergavenny I always have the sensation of going suddenly back in time to a former world. These are no domestic Cotswolds. The heights top eight hundred metres: Waun fach in the Black Mountains is 811 metres high; Pen-y-Fan in the Beacons 886 metres, almost on a par with Cader Idris. Yet all this is wastage from a range that once towered high over Wales. Some of the sandstones preserve evidence of powerful rivers, even of channels that cut down into underlying strata. There are the dumpings from floods, and slow settlings in lakes. Not all the rocks are red: the Brownstones speak for themselves (as do the Brown Clee Hills to the north); cornstones are odd, blotchy red-and-white sandstones. The rocks mostly dip only gently, so that the hillsides sketch out the sequence of the strata.

The topography of the Brecon Hills is more rounded than the peaks of the Cambrian mountains; the sculpting by the last Ice Age is still there, but muted, softened by erosion. Simpler structure is spelled out in the landscape, which is altogether open and bare. Steps in the hillsides, or crags, are usually the harder, thicker sandstones, and the eye can often follow the same bed for several miles. Gentler slopes between may be siltstone layers, where fossils are usually to be found. Where upland streams cross the sandstone benches they make charming, tumbling waterfalls cascading from step to step, seeking out any plane of weakness. The caps on Pen-y-Fan and Corn Ddu are patently geological: they are made of massive sandstones appropriately called the Plateau Beds, coming

---

*My colleagues who work upon minute fossil spores tell me that the fossil record of the kinds of spores that were likely to be dispersed in air goes much further back – even into the Ordovician. Perhaps there are early plants still to be discovered which will show an early move from the safety of water. I acknowledge here that Devonian spores are of the greatest use in correlating rocks deposited in one basin with those in the next. They have been responsible for putting the historical narrative of the 'Old Red' on a scientific footing.

near the top of the Old Red Sandstone, and announcing their geological position quite blatantly at the top of these hills; the caps act like hard-hats to protect the rocks beneath from further erosion. The Brownstones make gentler profiles. In the Usk Valley near Abergavenny there is a nearly conical hill, the Sugar Loaf, which I have heard described by an authoritative local as 'an extinct volcano'. It is not that, of course, but it *is* another 'Old Red' hill capped by an outlier of tough sandstone. A brisk walk to the top will allow you to look westwards towards the greater hills of the Black Mountains. Imagine now that where you stand was the bank of a recently flooded stream, and that towards the Black Mountains there were sandy braids and banks left by other floods, and pools, and perhaps a whisper of green from a patch of *Cooksonia*. And beyond that the towering Caledonides blotting out the tropical sun. For a moment you will be in the Devonian of south Britain, at the time of the greening of the Earth.

The 'Old Red' country of Hereford and Worcester is altogether gentler. Cattle fatten in water meadows, until ready for sale in one of the markets at Hereford, Leominster or Ledbury. Cattle bred on the poorer pastures of Wales are often 'brought on' here. Comfortable ancient farmhouses prove that affluence is no novelty here.

The Old Red Sandstone provides a ready source of blocky building stone in Brecon and Monmouth and Abergavenny, and anything monumental will likely be made of it. This stone can be a bit ponderous and empurpled, although it does have its own sense of place. But tiny Kilpeck Church in Herefordshire is a total delight, with some of the best-preserved and richest romanesque carving in our islands. The organic forms are oddly reminiscent of Devonian plants, as if the anonymous carver had somehow intuited form from the stone itself . . .

The story is completed by the other areas of Devonian south of the Bristol Channel, in the peninsula of the South-west: Somerset, Devon and Cornwall. This is where the other face of the Devonian is discovered, its dark side. This is where the rocks turn from red to black.

The northern belt of Devonian rocks extends from the coast, near Ilfracombe, eastwards to embrace Exmoor, and the Brendon Hills, and then further east again to appear as an inlier in the Quantock Hills. The southern belt includes all the westernmost part of Britain that is not otherwise made of igneous rocks: most of the south coast from Torquay to Penzance, and virtually all the north coast from Tintagel to St Ives Bay. This is the Cornwall of coves, and of wreckers and

The dark face of the Devonian: Dartmouth (right) and Kingswear Castles, Devon, built upon banks of cleaved Dartmouth slates.

smugglers, Arthurian legend, and Daphne du Maurier. Between the northern and southern belts there is a huge tract of overlying Carboniferous rocks – the Culm – thus proving that the whole of the South-west west of Lyme Bay is an enormous syncline.

The transformation in the colour of the rocks from the red of Herefordshire to the black of Devon is important. It is progress from Devonian freshwater and terrestrial environments in the north to a vanished Devonian ocean to the south. So complete is the transformation – so utterly different the rocks – that for many years the time equivalence or not of the Old Red Sandstone and the marine Devonian of Devon and Cornwall was a matter of the most vigorous disagreement. Martin Rudwick has dissected the history of this hot controversy of the nineteenth century in a scholarly and entertaining way. There is more than a passing reminder of the battles over the Moine Thrust: indeed, some of the protagonists were the same. I wish to take from this just one concept, encapsulated in the word *facies*. This embodies the notion that different kinds of rocks have been deposited at the same time: thus the Old Red Sandstone is a non-marine facies of the Devonian. Once stated, it seems obvious. But then, there is often nothing more important than a concept which is obviously true, especially when acknowledgement of that truth has been hard won.

The dark Devonian was laid down within a southern ocean, which has now vanished just as Iapetus did before it.

Devon gave its name to this division of geological time. Only the most rabid chauvinist would claim that the development of these rocks in the South-west compares with any in the world; everybody else would admit that there are better and fuller rock sections in France and Germany. But near Torquay, and around Torbay, there are limestones which give at least a glimpse of the richness of life in the Devonian sea. There are fossil corals, which form reefs. There are some trilobites, including the famous *Phacops*, a trilobite with vision so perfect that a photograph was taken through one of its lenses after more than 360 million years entombed in the rock. There are brachiopods, and early ammonites, which are just taking up their role as matchless time-keepers for the geological column. The limestones are often impure, which makes them coloured. Around Chudleigh they are often reddish. But other, dark limestones bear the coral *Thamnopora* outlined as a white tracery on the blackness, delicate as fishnet. It provides a wonderful facing-stone when it is cut and polished, and the Victorians loved it for their richly patterned fire-surrounds.

It must be admitted that the inland scenery which is underlain by Devonian rocks, lying west of the granite uplands of Dartmoor, is not generally remarkable. An uplifted plateau, it has been often compared with the south-western tip of Wales near St David's. Like that area, it has been anciently occupied, as proved by the standing stones and burial mounds. It can be bleak, and the wonderful place-names, Praze-an-Beeble, Tregarillion, promise an exoticism which is not there, and who were the saints that are celebrated only here – St Kew, St Just, St Tudy, St Mabyn and St Veep?

But the coast is a marvel. The sea, the open Atlantic which can be so implacable in winter, is now finding every weakness, seeking out every fault at the edge of the plateau. After such scrubbing and pummelling only sand remains, and when the sea subsides there are sandy beaches flooring coves. This is white sand, the durable, silica residue from igneous rocks like granite and from sedimentary grits; in places, rarer, even precious minerals can be found –

tourmaline or olivine. Surely these beaches cannot be bettered for young natural historians.

The waters of the North Atlantic are rich in nutrients, and marine life is abundant. Erosion leaves rocks standing from the sand like groynes, and around the rocks – and on them – there are rock pools full of seaweed, anemones, algae, little fish, crabs, snails: each pool an ecosystem refreshed by the tides. Barnacles and limpets and periwinkles live on more exposed rocks. The perfect summer day is spent prodding and digging and seeing what a net will catch; when you see how children are perfectly absorbed in exploring and pottering, you begin to believe in theories which would have Mankind evolve upon a beach.

But look towards the cliffs. There will be dark rocks; more likely than not they will be slates or grits. There may be obvious folding, or a fault eaten out by the sea into a cave. You immediately know: *these rocks have suffered*. Like the slates in North Wales, the black Devonian sedimentary rocks have been caught in powerful earth movements, which have squeezed and distorted them, often imposing a cleavage which makes the rocks break naturally in a way which does not correspond to the original sea floor. This is why, so often, the grain of the rocks is vertical, and why variations in their hardness readily produces bays, and ramps, groynes and stacks. The geology, as it often does, has dictated the ways in which erosion could sculpt the seascape. These dark rocks were squeezed in the Hercynian mountain building phase, which folded the Devonian and Carboniferous of the western peninsula in complex ways which geologists are only now beginning to unravel. Where streams have followed flaws and faults little coves and inlets are the result, often making natural harbours, like Mevagissey or Gorran Haven. There are villages of rough-and-ready, but almost impossibly picturesque cottages. More extensive 'drowning' of river valleys caused by the rise in sea level following the melting of Ice Age glaciers, has produced the dramatically deep indentations of the coastline at Plymouth Sound and the harbour at Fowey.

Black slates suitable for roofing have been excavated in great quantity. The Pengelly quarry at Delabole in north Devon is possibly the only runner-up to the Penrhyn chasm in the Cambrian slate belt. Slate does everything around Delabole: builds stone walls, often in a herringbone pattern; roofs houses (as over much of Devon and Cornwall); strange little cottages have hung-slates making their walls.

There is something eerie about this part of the Devon coast, between Port

Isaac and Tintagel. Little streams cut steep-sided valleys – almost gorges – which run daringly to the sea; some tiny havens have just one road in and out, giving a certain claustrophobic atmosphere. Possibly because the coast faces northwards, the poorly lit darkness of the rocks along the plunging cliffs seems portentous and sinister, as if inviting wrecks. Antique remains of fortifications brood against the turmoil of the sea. The ruins of Tintagel Castle conform to every ideal of the Tolkienesque kind of castle, perched recklessly above the sea as if to impress by vertigo alone. For once, an invocation of Arthurian romance does not seem sentimental exaggeration; if Merlin's ghost were to prowl the cliffs anywhere, it would be here.

Slates are not kind to fossils, as *Angelina* demonstrated in North Wales. The dark Devonian does not commonly yield evidence of past life. But even in Delabole quarry there are traces sufficient to show that these slates are not the same as those that top mountains in Wales or form slithery screes in the Lake District. The 'Delabole butterfly' is a spiriferid brachiopod which when crushed and stretched in dark slate might pass muster for some lepidopteran crudely sketched on the rock. To an expert eye, it is related to perfect specimens which can be recovered from Devonian limestones in many parts of the world, and indicates its antiquity almost as if millions of years had been engraved like a label in the rock.

The Lizard should be mentioned here, even though it does not properly belong. It, too, was caught up in the Hercynian vice, but it is probably a lot older. It is exotic with its weird, green serpentine rocks around Kynance Cove, and dark igneous and metamorphic rocks elsewhere; on this peninsula there are no ancient sediments. It *feels* like an antique world, inducing the same sensation as when you climb on to the Longmynd from the valley at Church Stretton, and the Lizard has indeed been claimed as a piece of Precambrian 'basement', thrust up from the depths. It is home to a special flora. The Cornish Heath (*Erica cornubiensis*) abounds on the weathered serpentine, which produces exactly the right kind of slightly alkaline soil.

There is a frontier between the two complexions of the Devonian. This is Exmoor, the Brendon Hills, and the Quantocks, where the Devonian rocks come to the surface in an east-west belt fifty miles across, lying north of the Carboniferous rocks of central Devon. Here there are sandstones, and also darker silts and shales. In this region, the marine influences of the black Devonian to the south and the terrestrial and freshwater of the 'Old Red' to the north oscillate and

St Michael's Mount, Marazion, Cornwall, at half-tide in June 1906, the most spectacular of several St Michael's — dedicated to the slayer of devils. It is built upon igneous rocks that have ensured its endurance in the face of erosion by the sea.

alternate. There are deposits laid down in ancient estuaries and deltas; they share Hercynian trauma with the dark rocks to the west — so that they are usually steeply dipping, and often folded. Those with strong nerves can examine the rocks in cliffs near Baggy Point or from the coastal path. Or geology can be combined with beachcombing near Woolacombe. Those who prefer a gentler walk can take the path south-eastwards from Lynmouth along the valley of the River East Lyn. On a fine day it would be hard to better this valley. The stream bounds from boulder to boulder. Trees line the valley, but not so densely as to obscure sunny ledges, where bilberries can be found. If you are lucky enough to find some, you cannot keep it secret; the purple fruits always stain the lips like a cosmetic. Lichens and polypody ferns decorate older branches of oak trees; this is epiphyte country, where warm rain encourages plants to grow wherever there is a purchase in the cleft of a tree or on a fallen branch. But huge boulders in the stream bed are a reminder that in spate the musical Lyn becomes a torrent which can, literally, move mountains. One real flood can transport more than a century of regular erosion.

Many other Exmoor rivers drain southwards, and they are as delightful. I doubt whether there is any valley in the southern half of England which gives more pleasure to the walker than the upper course of the River Barle, weaving back and forth through valley copse and hillside from Dulverton on its way to Withypool. Some routes go back through centuries. There are ancient packhorse bridges built from a patchwork of sandstone blocks, and older houses are stone-built in the same way, with or without whitewash. Open moorland between sequestered valleys is typical ling moor developed on peaty soils, as near Dunkery Beacon, and shows what the moors must have been like before farming made new inroads upon the open country. These newer farms are surrounded by high beech hedges, which line the twisting lanes so that in places you feel that you might be entering a maze.

It was in Dulverton that I ate the perfect cream tea. There is no greater sensual indulgence: cream so thick it is reluctant to leave the spoon, strawberry jam heavy with fruit, and crumbly scones with a hint of astringency to balance the sugar in the jam, and all piled up as high as they will go. Although it is *possible* that the cream originated on the New Red Sandstone in mid-Devon, I prefer to believe that it was produced by cows grazing water meadows flanking the lower reaches of the nearby River Barle or the River Exe. Thus, by the migration of indestructible molecules, the waste from the Caledonian mountains shifted across plains towards the Devonian sea and was incorporated at last into the 'Old Red'. The Hercynian movements folded these rocks, which were then eroded deeply. Ultimately, they came to supply nutrients to the grass grazed by cows that produced the cream – cream which was the centrepiece of the perfect cream tea. So it was that molecules from Caledonia itself came to be eaten with considerable pleasure in Dulverton in the late twentieth century.

9

Fells and Dales

CORAL REEFS have a certain glamour. They have a richness about them, a tropical plenty: a hundred fish of every colour, crabs, lobsters, sea anemones, and corals themselves, branched and horn-like, or massively obdurate, some convoluted like the lobes of the brain, others spread like flowers. The sea water is always warm and clear, rushing at the reef-flat in a cloud of foam, where, making a virtue of toughness, specialised organisms like coralline algae and sponges have learned to thrive. Reefs are prolific and apparently chaotic places, yet fragile, too. Pollution threatens them; or a change in salinity or temperature, or the influx of sediment. Once the coral fabric dies the whole thing degrades, until only a rotting tower of dead coral remains, a hulk, virtually lifeless.

Reef exuberance creates rock. Coral rock – a limestone, naturally – combines with all the debris tumbled from the reef and all the shells of creatures that lived in the reef. And this rock can build mountains. This is the limestone that stiffens the spine of England.

After the Old Red Sandstone was deposited, the sea returned over central Britain and to Ireland; not all at once, nor yet completely, because the destruction of the Caledonian Mountains would take much longer (an elevated area known as St George's Land persisted through central Wales, for example), but sufficient for shallow and warm seas to spread progressively northwards. This sea laid down sediments upon the eroded remnants of much that was older – even directly upon the Precambrian 'basement' in Anglesey. The rocks that resulted from this great *transgression* of the sea are known as the Carboniferous Limestone. This formation records one time in our history when the British Isles gloried in reefs, when sun and clear seas combined to nurture reefs in places which are now high above coal towns, and where a sheep may clatter over a head of coral that grew towards the light in the sea three hundred and fifty million years before.

There is something of a contradiction in the name. Carboniferous implies 'yielding carbon or coal' just as fossiliferous implies 'yielding fossils' or metalliferous 'yielding metals'. Limestone does not, of course, yield coal. However, the Coal Measures that follow in the geological column *are* the source of coal in Britain and in many other parts of the world – the typical 'carbon yielder'. In naming the geological systems both Pennine limestones and the rocks of the coal basins (together with the 'Millstone Grit') were shackled together under a single flag – the Carboniferous System. This marriage is not an altogether happy one, which is one of the reasons why the North Americans divide this time into two systems which they term Mississippian and Pennsylvanian, respectively.

The Carboniferous Limestone is the first great sculptural contribution that limestone makes to the British landscape. Two more were to follow: the Jurassic limestones and the Chalk. Limestones stand high enough to define the skeleton of much of the landscape outside Caledonia and Wales; between them, these three limestone formations lay down the ground rules, drawing vale from upland, directing river and road. Industry and large towns particularly avoid the Carboniferous Limestone. It has a propensity for wildness. This is not least because it is tough stuff, the hardest of the limestone formations. Even though rocks containing corals would have started as quite porous and full of cavities, the passing of geological time has filled every pore with calcite, a cement that binds and renders it dense. Parts of the limestone are massive, hardly bedded, and resistant to the worst that weather can do, with something of the obduracy of granite.

The Pennine chain is the central outcrop in England, running north from Ribblesdale, through Yorkshire, Durham and Northumberland, to the East Coast, centred on Holy Island. It continues southwards as moorland underpinned by Millstone Grit. Try though one might to avoid the cliché, the Pennines really *are* the backbone of England, although scarcely the stiff one that another cliché always compares with a ramrod, because it is a disjointed and shuffled backbone, and a branch swings off almost to encircle the Lake District, with the Vale of Eden interpolated on younger rocks lying to the west. The Lake District would have stood out as an island of Caledonian pedigree, ancient even then, as the warm, calcareous seas of the Carboniferous lapped its shores.

To the north, in the Midland Valley of Scotland, girdles of Carboniferous Limestone encircle the coalfields, proving that limestone rocks must also pass beneath these coal basins. Southwards, the limestone of the Peak District of Derbyshire breaks through the cover of younger Carboniferous rocks at the centre of a broad anticline which separates the Lancashire Coalfield from the Yorks-Derby-Notts Coalfield, to either side of the midriff of England. Imagine a great Hercynian arch bent north–south through which the deeper strata appear at the centre, like entrails in a carcass. In the South Wales coalfield, however, older Carboniferous Limestone embraces and encloses the younger coal-bearing rocks from Pontypool to Llanelli in a huge basin. On the south side of the coalfield it makes up the Gower Peninsula, which stands out emphatically against the sea on the east side of Carmarthen Bay, on the west side of which it continues further, folded more intensely, between Tenby and St Govan's Head. In North Wales, a north-south strip of limestone lies to the west of the Flint coalfield, comprising the fine Eglwyseg Mountain, near Llangollen. There is a stretch of Carboniferous Limestone centred on the Wye Valley Gorge north of the River Severn, and south of it around the Forest of Dean, and in the most famous inlier of all, the Avon Gorge, where you can look down upon it from Isambard Kingdom Brunel's suspension bridge at the edge of the city of Bristol; thence into Somerset in the east-west strip of the Mendip Hills, which includes Cheddar Gorge.

But the greatest area of Carboniferous Limestone underlies the whole central plain of Ireland. You can drive from Dublin to Galway Bay and The Burren in the west without leaving it, and then you can take a boat to the Isles of Aran and see it further west again, horizontally disposed, in cliffs ranged against the open Atlantic Ocean. North to Donegal Bay and south to Kilkenny the clear Carboniferous sea bathed the land. The older, Caledonian history only peeps through in inliers, as if the earlier history were being hushed up. Nearly horizontal limestone smothers the earlier story over much of this country. Towards County Cork and Kerry in the South-west more intense Hercynian folding around Dingle Bay can be traced across St George's Channel from South-west Wales.

Carboniferous Limestone covers more of the British Isles than any other formation. Its pale blue on the geological map recalls its shallow blue seas, sapphire blue when there were reflections from white lime sands forming shoals in lagoons, coerulean blue over deeper stretches of sea. The Hercynian convulsions have disturbed its original distribution, but it is still much easier to draw together in the mind the patches of outcrop remaining into a shimmering stretch of blueness than to imagine great and vanished Iapetus. The corals are still there, just as they grew, or in blocks tumbled from the reef-front. Their colour has to be imagined, amid the clamour of life that always accompanies a reef, but this is not to stretch the imagination too far . . . There are corals that look like organ-pipes, others are massive hexagons pressed together like the basalt in the Giant's Causeway, yet more like fat cigars, or like fine honeycomb. Now, they are usually bleached pale grey, painted out by the British climate. You may have to look closely to see them, but they will betray their presence because of their network of septa, which were, in life, part of the supporting struts for the coral polyp. They will look like delicate spider's webs, little orbs or wheels set about with a filigree of spokes. The wonder is how clear the corals become if the rock is polished. Their calcite is usually white, whiter than the body of the limestone, so that they become drawn on the rock surface, down to the last septum, with the kind of clarity that classical anatomists used to sketch out the nervous system in the body.

Now, having painted a picture of endless coral, I have to spoil it. There is less coral reef in the Carboniferous Limestone than I have implied, although it is important in many places, and supplied much debris to comparatively mundane rocks. The reefs grew particularly towards the edges of upstanding blocks at Craven and at Askrigg, ramps almost, the edges of which were defined by faults. Many of the more massive Carboniferous limestones are rather intransigent rocks made from tiny, broken bits of fossils. In basins beyond the reef 'blocks' there are shales and limestones alternating endlessly, or, in the North Pennines and Northumberland, sandstones, limestones and shales, arranged on top of one another like layers of Neapolitan ice cream. All these are examples of different contemporaneous facies, like the red and the black aspects of the Devonian. Beneath the clear Carboniferous sea there were reefs in places where the water was clear, lagoons in other sites, while elsewhere in the deeper, fault-bounded basins, shales accumulated and ammonites preferred to swim. The simple, blue colour may be apposite, but it covers an unsignified richness. Like so many matters geological, the closer the observer looks the more he realises that the world in the past was every bit as full of nooks and irregularities as it is now. The Carboniferous was a varied tropical seascape.

One special, but not uncommon limestone is composed of the remains of the stems of sea lilies (crinoids). The stems usually come in short, pipe-like sections a centimetre or so across and several centimetres long. End on, they look like coins, or, having a perforation in the centre, like flat beads that might be strung on a necklace. Jumbled together, they make an appealing pattern, which has been polished to advantage to make an ornamental 'marble'. This facing-stone was used generously inside the Royal Festival Hall, on walls in the foyer. When I first examined these crinoidal stones with my hand lens, going up rather close and peering, slightly bent, I found myself surrounded by a dozen curious onlookers.

It is the massive limestones, however, which dominate the high tops of the Pennines and The Peak. They make the bare summits; the resistant, nearly horizontal beds are the fabric of the bluffs which stand out like the walls of ruined fortresses among the trees in the dales, and the vagaries of their cliffs pose interesting problems to the rock climber. The alternation of fell and dale defines the landscape; rivers biting deeply into the Pennines from the east make sharp saw-cuts into the intractable and massive limestones. Those who have followed the Pennine Way will confirm that it is altogether more up fell than down dale. The cuts made by the Rivers Swale, Ure, Nidd, and Wharfe still dictate absolutely the passage of roads and the siting of villages. And where a lesser dale joins a greater one, there, too, a minor track or road will follow. And from the smaller streams still smaller becks and gills draw the course of tiny valleys, or run steeply down from the top of the nearest fell. There are often roads on *both* sides of the dales, and the driver may thus go in and out of the dale along a subtly different route and be set the pleasurable task of deciding which is the more attractive. But between dales there are only tiny roads with hard gradients as they take the plunge down into the dale at either end. From Wensleydale, for example, two very minor roads cut across Askrigg Common to Swaledale, running easily and open to the common at the top, but swooping downwards in sight of Swale and Ure. Incidentally, there is no River 'Wensley' to accompany Wensleydale, and if logic were followed – which in England it seldom is – it ought by rights to be called Uredale.

At the moortops, obviously overlying the Carboniferous Limestone there are frequent patches of Millstone Grit, which shows up in the darker stone used in the construction of the dry stone walls. In places, one could almost map the geology by wall colour alone. At Ingleborough, the Grit supplies a flat-topped summit, which looks curiously foreign to the waste of limestone all around. Because the Grit produces acidic moorland, while the limestone is associated with calcareous soils, the whole aspect of the plant life changes in tandem with the rocks.

Above   *Ingleborough rises above the plateau of the Great Scar Limestone (Carboniferous Limetone), its slopes composed of Yoredale rocks, its cap of the resistant Millstone Grit.*   Below   *Carboniferous Limestone pavement at Malham in the Pennines: specialised plants nestle in miniature gardens in the 'clints'.*

Escarpments are 'scars' in this Carboniferous Limestone country, and what better term for such exposure? For here the anatomy of the land is truly displayed; a wound, a gouge, no modesty in exposure of its innards. The Great Scar, east of Settle, gives its name to a limestone formation that makes many of the features in this part of the Pennines, where the rocks are close to horizontal. Gordale Scar is darkest of all, a grim funnel carved northwards at the head of the River Aire. The romantic painter James Ward caught its brooding quality in his famous painting *Goredale Scar*, although perhaps there is almost *too* much drama in the scene, as it verges on the bombastic.

The Carboniferous Limestone is one formation that defies concealment. All through the Craven District, and above Horton in Ribblesdale, and north-west of Ingleton, in Scales Moor and Raven Scar, the rocks are naked on the heights, and no covering of soil endures. The same can be seen in Derbyshire, as at the marvel stones, east of Smalldale, or in the Burren in western Ireland. These are limestone pavements although, fissured and dissected as they are, they resemble paving only from the sky. Close to, the fantastical patterns created by weathering look like a relief model of mountain ranges and gorges from another land: a hidden landscape within the landscape. You can lie still, imagining yourself a traveller in this landscape, weaving your way through passes between fretted ranges, or avoiding grim gulches. This miniature scenery is the work of rain upon limestone, which alone can discover the weaknesses in such grey and intractable rocks.

The fissures are known as clints or grykes. They have generally developed along joints, hairline cracks developed in sets at right angles, vertical to the horizontal bedding planes. These cracks are wide enough to allow rainwater to squeeze in. What opens up the cracks, and what eventually turns the limestone pavement into its crazy corrugations, is a chemical fact: that rainwater can dissolve the very rock, a few molecules at a time. Rainwater absorbs a small quantity of carbon dioxide from the atmosphere to produce a very weak solution of carbonic acid, and it is this acid which gets to work on the calcium carbonate, the body of the limestone, taking away minute quantities in solution. Time, then, and more time, the summation of a million rainstorms, and there will be a fluting of the surfaces, a slow rounding, a crack deepened to become a gulley. And so with age the limestone face becomes pitted and grooved, until, as with all ageing, its character becomes etched into its physiognomy.

Nor is this the end of it. For the water percolates downwards through the rock, dissolving and widening fissures, until there are galleries of fissures snaking their way through mountains, and extending far below the ground. And then, when such a widening crack catches a stream from the high ground, it, too, will plunge into the limestone, twisting and turning through the rock, until it emerges again at a lower level, released from its involuntary incarceration in the underground. This is how systems of caves arise, and potholes, and narrow passages, and the whole collection of subterranean tortures favoured by the most masochistic outdoor enthusiasts, speleologists. One shudders at their descriptions of wriggling through passages in total darkness, like a maggot through an apple, often immersed in murky, icy water, deeper and deeper in the ground and with no guarantee of an exit if there happens to be unexpected rain ten miles away. There is, however, no landscape more hidden, more arcane, than that of the underworld of caves, and only the Carboniferous Limestone produces extensive cave systems in Britain.

There is something compelling about caves. It was formerly possible to walk into some of the caves high up on the sides of Cheddar Gorge, and who could resist peering into the blackness and getting a delicious whiff of Hades? And perhaps there is something of Persephone's return when a stream bubbles up again, or rushes out of its underground entombment, cold and effervescent. In the southern Pennines, the stream draining Malham Tarn disappears underground a short distance south of the tarn only to emerge again at the base of the limestone precipice, far below, at Malham Cove. It is the same water, no doubt about it, because dyes thrown in above, well up below. At Porth-yr-Ogof (Brecon) the Avon Mettle comes out of blackness pure and cold. Even on the tourist trail at Wookey Hole on the east end of the Mendips the chattering voices of the visitors become hushed as they contemplate the nascent River Axe within the artificially illuminated caves. The whole of Cheddar Gorge is a collapsed cave system, but where an underground river used to run, there is now only the road. At levels deeper and more inaccessible water continues to cascade through widened joints or flow more gently through caverns which few eyes have seen.

Early inhabitants of our islands used these Carboniferous caves to shelter from the cold, to butcher meat, to perform magic. Now, the tools that they used, the bones of the animals they ate and those that ate *them*, are preserved in layers on the cave floor. There are cave bears, deer, wolverines, and voles. They may be sealed in stalagmite, which is a limy deposit produced by *reversing* the process which rendered limestone in solution. Recently, methods for radiometric dating of stalagmite have improved the chronology of such deposits. They are widespread: in Wookey Hole and the caves along the Mendips; in the Peak District, the caves of Creswell Crags, including the irresistibly named Mother Grundy's Parlour; on the Gower coast Minchin Hole and Bacon Hole, once seaside caves but now stranded above the hungry waves as a result of changes in relative sea level after the last Ice Age. Seven thousand years ago or more human occupants of Carboniferous Limestone caves would have been as nervous as any of the animals they hunted, not yet masters of the world, grateful for the solidity of the rock that protected them. Could they have failed to be awestruck by the darkness of the deeper caves, concocting trolls or hobgoblins from imagination, building myths which maybe unconsciously linger with us even now?

It is one of the strange connections of the hidden landscape that the clear seas of the early Carboniferous came to house pioneer inhabitants of these islands, living on the edge of a darkness where even troglodytes would not venture. 350 million years had passed. No massive limestones and no caves would have been possible without Britain having been in the tropical belt where calcium carbonate could accumulate in great quantity.

There are odd manifestations of differential erosion of Carboniferous Limestone. A small road runs northwards from Hawes in Wensleydale to Muker in Swaledale across Abbotside Common. A short distance to the south of a hill called Hood Rigg the road turns down into a sharp-sided valley, a typical beckside. On the west side of the road there are the Buttertubs. It is as if some demented engineer had driven down not one, but a whole succession of wells into the bedrock in a fruitless search for water. Some round wells were abandoned quickly, others pursued too deep to plumb. Ferns decorate their rims. The tubs are no more than the record of waterworn breaching of a resistant layer into softer horizons below: the kind of natural phenomenon that looks persistently unnatural. Seven miles to the north, on the far side of Rogan's Seat, the Tan Hill Inn claims to be the highest in England, and to Pennine walkers the most welcome.

Thick Carboniferous limestones have been much exploited: quarried for building stone; burned for quicklime and cement; beaten into chips for roadstone and railway ballast; used for flux in smelting; boiled away in the chemical industries. Quarries on a mighty scale have been opened in hillsides, making scars by any judgement. Many are still working: near Cracoe in the Pennines, in Derbyshire, near Crickhowell on the edge of the South Wales coalfield. But those that are not, stay obstinately pristine for a long time. Sheer, grey walls of limestone have

Dove Holes, along the Dove Valley, Derbyshire, a modest example of caves eaten into Carboniferous Limestone by the corrosive action of water. Such caves were an important refuge for early humans in our islands.

not acquired the mottling and subtle relief which add interest to natural cliffs. There can be a conflict between the geologist's desire to continue to see what is happening in the rocks and those who wish to minimise environmental 'damage'. To many observers, quarries are little more than eyesores, but others will see them as a window into the past. One could imagine the paradox of a preservation order being placed on a quarry because of some unique geological feature, while another conservationist lobby would seek to heal the 'scar' and restore the landscape. In defence of the quarry, as Nature blunts the worst excesses of the quarryman, plants and trees return, water collects, and then quarries tend to present a variety of habitats which are becoming unusual – comparatively unpolluted ponds, for example. A little sympathetic restoration should enable both the geology to be preserved, *and* the habitats to be enhanced in the long term.

Nobody would object to the use of Carboniferous Limestone in buildings, or stone walls. The first feature that strikes the visitor to the dales is an endless succession of stone walls, running along the roadsides, bounding fields near the stream and snaking up hillsides – even surprisingly steep ones – towards the open tops. They are constructed from stones of various shapes and sizes, plucked from the fields, and apparently miraculously bound together in a three dimensional jigsaw, topped with a line of stones arranged cross-wise. Farm boundary walls may approach six feet high, but those between fields are a little more than four feet high, enough to stop a sheep jumping over. Some walls have 'creep holes' specifically to permit the passage of sheep between fields. The shrewd observer will note that these are too small for most modern sheep, which is a reflection on modern breeding techniques. The arrangement of field boundaries is that arrived at after the Enclosures, but there are hints preserved as subtle terraces on hillsides of truly ancient field systems, possibly dating back to the Iron Age. It is certain that Norse settlers were there in the dark ages.

But the Yorkshire Dales have been sheep country for centuries, prosperous and well-watered, with farmsteads dotted quite regularly along the valley sides. If an example were needed to demonstrate harmony between rock foundation and human habitation Garsdale and Arkengarthdale would supply it. Even the most ardent wilderness lover must admit that the buildings and walls only add to the landscape, providing a scale for perspective that nature alone does not supply. Spring light reveals a kind of pearliness in the blocky limestone walls of the farm buildings which gives the lie to the idea of grey as a dull colour. Then the different hues on roofs and walls make for a subtle patchwork of tones which mirrors the exquisite differences in shades of green between one field and the next. The farms are often set about with groves of sycamore or ash. Two storey farmhouses may have their byres and barns attached, particularly the older ones. But field barns, or laiths, are elsewhere scattered through the fields, often with a porch on one side for driving in hay wagons. There are some fine examples near Malham. Stone roofs are the rule, often with a somewhat browner tone than the walls; the limestone 'slates' derive from the Yoredale rocks which overlie typical Mountain Limestone in parts of North Yorkshire. Many of the farms date from the seventeenth century, and with the endurance of limestone built into their fabric they should decay only as fast as stone weathers. Incidentally, in Arkengarthdale there is a hamlet called 'Booze'. Rather mysteriously, it is a mile away from the inn at Langthwaite.

The weaving industry in Yorkshire has left its trace in weavers' cottages, and these were also stone-built, with long rows of windows signalling the rooms where looms worked at the cottage industry. Weavers needed good light. Towards the end of the eighteenth century water-powered machine looms began to replace the home weaver, and then steam engines, fired with coal, escalated the technology a further notch. Ruins of chimney stacks – now often grotesquely overgrown with ash trees – occasionally catch your attention even in the remotest beck.

The potential for water power is demonstrated whenever the dales rivers take a sprint over rapids, tossing the long locks of water crowfoot (*Ranunculus*), and providing a favoured place for dippers. Water seeks out weaknesses; a subtle difference between one bed and the next will be worked upon by a stream until a stretch of broken water becomes a step, and a step becomes a waterfall. In Pennine country, where strata are nearly horizontal, sidestreams often join the dale in a staircase of broken water, often scarcely a trickle, but after rain, transformed into a zigzag torrent. Waterfalls are called 'forces' in the dales, and this is not an inappropriate name. There is a whole series of them in upper Swaledale, upstream from the gorge, near Keld: Wain Wath Force, Catrake Force; and Kisdon Force are among them. In upper Wensleydale, Hardraw Force is nearer the Pennine Way, and the most spectacular of waterfalls, dropping ninety feet into a turbulent pool full of boulders. It was possible to edge *behind* the waterfall, perched on a narrow and very slippery ledge of rock, not a bad place to reflect upon the influence the subtle differences of the Carboniferous sea floor still wield on the face of the landscape today.

Carboniferous Limestone stream water is pure, spring-fed, nimbly flowing. This is preferred by brown trout, but they do not often grow large in the upper reaches of limestone streams. Izaak Walton incomparably described fishing expeditions along the Derbyshire rivers Dove and Manifold in the *Compleat Angler*. Trout seem to enjoy lying close to the bank, a taste which has led to the invention of the technique of 'tickling'. It must be an esoteric skill; whenever I have tried it the fish have shot off at the first hint of hand in water.

Water from limestone springs will deposit its dissolved calcium carbonate over stones or other objects as a covering of crumbly, whitish *tufa*; this is a reversal of the process of solution which ate out grykes and caves. Petrifying springs at Matlock are famous for 'turning to stone' kettles, or boots, or almost anything hung up under the dripping water. It just takes time. Witches and warlocks who can turn people to stone add an horrific touch to myths and fairy-tales, but limestone seems a rather humdrum source for such potent terrors. I doubt whether the creator of the Medusa was inspired by calcareous springs even though they

Geology in control of waterfalls: Thornton Force, Kingsdale, Ingleton, North Yorkshire, where streams tumble over the massive Carboniferous Limestone.

are common enough in the Mediterranean region. *In extremis*, we are petrified, rigid with fear. I suspect that the sensation led to an analogy with rock, rather than any geological phenomenon inspiring the description.

Along Dovedale, there is a feathery-leaved and elegant parsley called sweet cicely (*Meum*), which grows in summer in those streamsides and waste places where quotidian cow parsley grows in the south. One might hide behind it in attempting to tickle a trout. From waterside to mountain top the Carboniferous Limestone flora is full of good things: a *bright* flora, if you like, and hardly sharing a species with that of nearby moors on the Millstone Grit, where heather and cotton-grass recall the wilderness of Scotland.

Many of the plants from dales and fells can be found on other British limestones – wild thyme, dog violet, field scabious. I shall return to these flowers again in describing the Chalk. The upland limestone flora often tucks

138

itself into hollows and cracks between the 'pavements'. Short fescue grasses set each hollow in a miniature garden, where violets and speedwells, bird's foot trefoil, barren strawberry and rock rose are all in scale. Occasionally, a stunted rowan tree will grow as a natural *bonsai* in the meagre soil. In the Burren (Co. Clare) 'pavements' spring gentian (*Gentiana verna*) and rare orchids, like the dense flowered, or the lady's tresses, grow commonly in the limestone gardens, but their rarity elsewhere is such that botanists make pilgrimages there from all over Britain. Towards late May The Burren is a mass of brilliant flowers – it could scarcely be more colourful if the whole area had been deliberately planted. Because the limestone adjoins the sea, there are cushions of marine flowers, too, like pink thrift or white sea campion. It is an extraordinary mix of alpine, Mediterranean and marine species; an ideal rockery, in fact. Our own attempts to reproduce the scene on patios in Penge are understandable but nugatory.

Even the driest limestone crevices of all still harbour a little collection of charming, small annuals which wither in the first hot days of summer: white-flowered vernal whitlow grass (*Erophila verna*), tiny, blue early forget-me-not (*Myosotis ramossisima*) and green, frilly parsley piert (*Aphanes arvensis*). Hutchinsia (*Hornungia petraea*) is probably as common in this habitat as in any other. Specialised lichens and mosses decorate the edges of such crevices, along with the tiny spleenwort fern (*Asplenium ruta-muraria*) which seems to prefer a diet of nothing at all. This flora is best examined with an ouzel's eye view on a warm spring day, shortly after rain.

For those with less exquisite tastes, or less myopic ones, the grasslands in the dales can be as joyful as alpine pastures, and there are high pastures named on maps in North Yorkshire. Buttercups and clovers, scabious and yellow rattle, knapweeds and hawkweeds provide the primary colours and there is every shade between. There are places where orchids are still quite abundant, mostly spotted and pyramidal orchids, but rare helleborines grow in protected sites in Derbyshire. That giant buttercup, the globe flower (*Trollius europaeus*), which bears spectacular golden goblets in late spring, seems to favour the edges of woods. Special Carboniferous Limestone rarities are listed in every British flora but one does not usually encounter them on a desultory stroll. The Cheddar pink is first among them; incredibly, it has been reduced in the wild in the Mendips because there are people who would rather dig it up than leave it be. The lady's slipper orchid is now little more than a legend.

But if one were to select a special Carboniferous Limestone plant which is not especially rare it would be the bloody cranesbill (*Geranium sanguineum*). It is a compact and orderly plant, with soft leaves like stubby hands, dotted over with almost perfectly circular carmine flowers which have a brilliance which seems somehow too exotic for our climate. If it had been discovered on some remote peak in the Andes it would have caused a horticultural sensation.

Ash woodland often fills the dales; ash is a lime-lover able to cope with shallow soils, and is therefore a natural coloniser. Ash trees are late in leaf, so that the woods stay lighter longer, and the canopy never achieves the overwhelming density of beechwoods such as those further south at Symond's Yat (Gloucestershire). Hazel trees establish themselves beneath the ash cover. It is a pleasure to walk through such airy woods at any time of the year. Wherever ash branches have fallen, it is impossible to miss seeing hard black growths the size and shape of potatoes distributed along the branches. These are the fruiting bodies of a fungus, *Daldinia concentrica*, cramp balls, which is an ash specialist. I have been unable to discover why these odd-looking cushions should have been considered as a specific for cramp. As for sympathetic magic, placed under a pillow they seem more likely to cause cramp than cure it.

There are other aspects of the Carboniferous. In Scotland, a sandier rock of the same age as the Carboniferous Limestone has generated the face of Edinburgh. It is a fine building stone, which can be cut to a good finish. It is responsible for much of the dignity and elegance of that city. There are not many cities with a centre so 'all of a piece', and which have been spared the worst excesses of tower blocks.

But in the middle of Edinburgh there rises a wild island among the genteel and respectable city, which might symbolise another side of the Scottish character. Gorse and bracken cling to the sides of an extinct Carboniferous volcano, Arthur's Seat. A sill, Salisbury Crags, lies to the west. Castle Rock is a volcanic plug which forms the crag at the very heart of the city, upon which sits the dour and intimidating fortification. The volcanic vents on Arthur's Seat have been called Lion's Head and Lion's Haunch, which is a bit fanciful. But it does not require much effort of the imagination to understand the volcanic *agglomerates* which flank the Seat, rubbly rocks which built the cone of the volcano; hardened through time, angular fragments now protrude through the rough grass. To the west, more extensive lavas now make the open hills of the Campsie Fells, north of Glasgow, and provide a ready escape into the wilderness from that city; these rocks are probably younger than those of Arthur's Seat, and contemporaneous with the deposition of Coal Measures. The Touch and Gargunnock Hills, west of Stirling, were part of the same lava field. One can visualise vents and volcanoes smoking above the lakes and the alluvial plains of the Carboniferous Midland Valley, occasionally galvanised into fiery fury. Arthur's Seat would have made a feature then, as now. The escarpment west of Stirling clearly shows terraces attributed to several different lava flows. The castle at Stirling actually incorporates a dolerite sill as part of its defences. Scottish Carboniferous volcanic rocks were made famous by the indefatigable Sir Archibald Geikie in his book *Ancient Volcanoes* (1897).

Probably the most famous Carboniferous igneous rock of them all is the Whin Sill which is late Carboniferous in age. The Sill extends for more than sixty miles across north-east England, with few breaks in outcrop, from near the headwaters of the South Tyne to the coast at the Farne Islands. It is *intruded* into the Carboniferous Limestone (which it must therefore postdate), interleaved among the bedding, as a sill should. The rock type is a singularly tough quartz-dolerite, 73 metres thick at its greatest. The word 'sill' originally

A great natural igneous rampart, the Whin Sill, is capped by Hadrian's Wall, Northumberland. Housteads Crags from Cuddy Crags.

meant a flat rock bed in this part of the country, while 'whin' meant hard, and so it is a strange coincidence that the name for this feature was possibly the same long before geologists recognised it for what it was. It has a dip and a scarp slope, and because of its toughness and resistance to weathering it makes a feature, with the steeper scarp slope facing north. It must have seemed like a gift from the Gods to the Roman emperor Hadrian seeking to wall off trouble to the north, a natural wall already and facing the right way. The Wall that carries his name is built on top of the Whin Sill for much of its length through Northumbria. It is still clear today, milecastles and all, especially near Crag Lough, north of Haltwhistle in the South Tyne valley; a section through the sill can be examined near the car park at Cawfields Crag. A Roman road dogs its course on the south of the Wall, while to the north one can imagine barbarians lurking in the Wark Forest. The Great Wall of China is massively larger, but exactly similar in structure: the patrolled wall, the barrier between the rule of law, and those who do not respect law; the army as defenders of business and empire. The wall is as much a line of flags, a signalling of the edge between ideas, as a physical barrier. Standing on Hadrian's Wall, the Whin Sill beneath you, it is impossible to miss the connection between the foundations in the landscape and the shape of history.

The Whin Sill peters out to sea in the Farne Islands, but to the ornithologist this is possibly the most important contribution that geology has to make to the British Isles. The grassy Inner Farne is a breeding colony for thousands of terns, which seem to thrive on proximity, as is the way of seabirds. If you land there, a furious, white blizzard of birds rises up, and the arctic terns are not above giving you a painful peck on the skull. On Staple Island there are puffins and guillemots. The Farne Islands may also claim to be Britain's, if not the world's, first nature reserve; St Cuthbert protected the eider ducks there. Eiderdown is, of course, a renewable crop; it is gathered from the nest lining and the harvest does not therefore interfere with the breeding.

St Cuthbert's Priory is on Holy Island, not far to the north-west. St Cuthbert no longer rests there; his bones were moved to Scotland after his death 1300 years ago, a journey irreverently described by Sir Walter Scott in *Marmion*. Holy Island is where the Carboniferous Limestone reaches the North Sea, as far away from its outcrop on the Burren, washed by the waters of Galway Bay, as it is possible to get. The island is cut off from the mainland by the sea for nearly half of each day. Seen from across the sands, Lindisfarne Castle is what every castle should be, turreted and remote, set upon a crag with the sea before. Limestone

Holy Island, where the Carboniferous Limestone reaches the North Sea.

outcrop forms the foreshore of the island; thanks to its resistance to erosion the island has acted like a natural groyne and gathered sands behind, which have made an area of such importance for wildlife that it is now a National Nature Reserve. Fossils of sea-lily (crinoid) stems are found on the south shore, and the reader will not be surprised to learn that they are known as 'St Cuthbert's Beads'. For some reason one detail of Lindisfarne stays in my memory, and it is neither a fossil nor an ornithological rarity. Behind sand dunes on the north of the island I saw more plants together of a saxifrage, grass-of-Parnassus (*Parnassia*), than I have ever seen. Each perfectly white flower is carried on an elegant stem, and even a few together are a pleasing sight, but here they were in their tens of thousands. I understood why the Muses would walk among this plant on Parnassus. I also knew the feeling that a Museum, dedicated to the Muses, should inspire but does too seldom.

10

Coal and Grit

IF THERE were a risk of sentimentality in tracing geological connections, a tendency to harp on about harmony between man and rock, then a brief look at the history of coal should act as a corrective. Much of the history of coalfields has been a story of oppression. It was in coal towns that the promise of the Industrial Revolution foundered on the inhumanity and greed of a thousand Gradgrinds. It is coal landscape that provides the quintessential image of pollution: slag heaps, sulphurous smoke creeping along mean terraces, ruined canals. Nor is this a nineteenth-century description. George Orwell's unflinching account in *The Road to Wigan Pier* is a report from the Thirties: ' . . . in the industrial areas one always feels that the smoke and filth must go on for ever and that no part of the earth's surface can escape them. In a crowded, dirty little country like ours one takes defilement for granted. Slag heaps and chimneys seem a more normal, probable landscape than grass or trees, and even in the depths of the country when you drive your fork into the ground you half expect to lever up a broken bottle or a rusty can.'

In my childhood, London was still heated by open coal fires, and every winter there were thick smogs, or 'pea soupers' (the yellow, split-pea soup, not the freshly green) which would provoke bronchitis and asthma – and even death. I was reminded of how things used to be in the industrial city of Tianjin in northern China. Every house had a crazy little chimney leaning out and belching a stream of creamy smoke. There were masks over the noses of the cyclists as they swarmed past in their thousands, braided like another stream, before they disappeared into the yellowing miasma. There was a catch in the throat, a sharp taste, a compulsion to cough, which suddenly and vividly recalled my London childhood: the sound of dry, persistent coughs coming through the smog, street lamps blurred and ill-defined and mentholated sweets which did no good at all.

The sharp taste of smog is caused by sulphur dioxide; this gas is released from coal when it is burned. Like carbon, which is the useful body of coal itself, sulphur is a product of the way coal was formed in the swamps of the later Carboniferous. Sulphur dioxide and carbon dioxide are released through combustion when these trapped elements react with the oxygen in air. Like escaped prisoners, they go on the run. Sulphur dioxide has been identified as the prime cause of acid rain. After many years of prevarication acid rain is accepted as a fact, and a catalogue of its depredations is no less serious for being acknowledged. Carbon dioxide is the 'greenhouse gas', which may be responsible for global warming, although there is no consensus yet on how much or how fast. In short, coal burning could be described as the most disastrous exploitation of a geological resource: when time had buried carbon in deep seams, was it wise so to exhume it and consign it to flames?

But one can scarcely blame Carboniferous trees for the problems engendered by industrial society, when greed is the culprit. It would also be disingenuous to pretend that *all* the products of burning fossil fuel were both bad and dispensible, as if, somehow, society could have been arrested at some pre-industrial stage, and all the incidental evils of industrial exploitation averted. This would have meant no steam, no steam engines, no iron or steel, no coal-tar chemicals. It is impossible to deny the contribution of coal power to the civilised world, and what is needed now are controls on pollution in the future, not lamentations over the past.

Coal makes up only a tiny proportion of the thickness of the Coal Measures. When you see the rocks as they crop out on the coast, as you may in South Wales on the east side of Carmarthen Bay, at Saundersfoot, or along the Firth of Forth in the Midland Valley of Scotland, the overwhelming impression is of sandstones and shales – often rather dark shales, to be sure, but scarcely coal. On the surfaces of some of these shales there may be impressions of the fronds of fern-like plants.

These impressions are themselves often 'coalified' – the tissues of the plant have changed to the element carbon. This happens as part of the fossilisation process. A stray frond settles into sediment – on the bed of a lake, perhaps – and becomes entombed within the sediment as more mud settles on top. Leaves and fronds are little more than collections of carbonaceous chemicals, forged in sunlight; as bacteria get to work on the frond many of these are consumed or changed. Burial accelerates the process and induces further change. Volatile chemicals may be driven off to accumulate elsewhere; when this process acts on vast numbers of single-celled plants this is the basis for the generation of oil. Ultimately, only a carbon film remains to record the frond, but this fossil may still faithfully show details of its surface. One of the first fossils I found in the Carboniferous of west Wales was a frond of *Neuropteris*. It looked rather like the frond of the living Royal Fern (*Osmunda regalis*), which is not uncommon in moorland country. In fact, it is not a particularly close relative of this plant, which shows that leaves and fronds must be used cautiously in identifying biological relationships.

Coal required special conditions for its formation. Waterlogged wood and other plant material accumulated in forest swamps in the coal basin. By the later Carboniferous forests almost as luxurious as those of the Amazon basin had spread widely over deltas and rain-soaked lowlands. The trunks of trees like *Lepidodendron* were as big as many forest trees today, and provided a good raw material for high quality coal. The trunks had to be entombed in the absence of oxygen, which under normal conditions encourages the breakdown of wood by fungi and micro-organisms, and ultimately guarantees its return to the soil. In a sense, the carbon in coal cheated death. The same *anaerobic* conditions encouraged sulphide minerals, and thereby also incarcerated acid rain. The potential coal seam was then sealed in by sediments above (which usually served to kill the trees as well) and, much later, after the heat and compression imposed by deep burial, the creation of coal was brought to completion.

Coal is restricted to seams, usually a foot or so thick, which must be followed by the miner into the heart of the earth; seams do not go on forever – they even split into two if a sandstone wedge appears. But in their pursuit of coal, miners follow productive measures far down beneath the cover of younger rocks, as under the Permian and Triassic which takes the ground between Nottingham and the Vale of York. The Kent coalfield is entirely concealed beneath the young rocks of the Weald. Because coal basins were under stress, and were also affected by the Hercynian folding and faulting, the course of the seam is not infrequently interrupted by a fault, when it must be pursued to its new level, or the search for coal abandoned. These faults are just one of the things that makes coal-mining so dangerous: one does not need to linger on the horrors of exploding methane or suffocating carbon dioxide gas, or the sudden collapse of the gallery roof in a shower of sandstone. To anybody who has felt, even fleetingly, the panic of claustrophobia the whole notion of mining coal, cramped and deep in the ground, befogged in choking dust, and deafened by the percussion of drills, seems terrifying. This tough life explains the extraordinary solidarity of the mining communities, a solidarity which has been further annealed through facing and surviving shared disaster.

Coal Measures were deposited over an enormous area, extending beyond our shores into France and Belgium, Westphalia, Silesia and Donetz. Disjunct basins remain: the Midland Valley of Scotland; east of the Pennines, the Northumberland and Durham in the north, and the Yorks, Derby, Notts coalfield to the south; the Lancashire coalfield to the west of the Pennines, and the Leicester field in the heart of the Midlands; the great South Wales coalfield at the centre of the syncline running east-west across Glamorgan; the Kent coalfield. Many a lesser field has been mined as long: the Bristol coalfield, Flintshire, Shropshire, or the Forest of Dean.

The Free Miners of the Forest of Dean were no more than family businesses exploiting their commoner's rights to mine. The hummocky landscape around Coleford (Gloucs) is the legacy of waste tips of these small concerns. Such ancient industrial exploitation as in the Forest of Dean leads to ambiguous landscapes. Certain coal mines, iron workings, or the pits of charcoal burners (900 of these were operating in the Forest in 1282) are so reclaimed by time as to leave the visitor wondering whether a trench or a pond discovered in woodland is natural, or a disguised legacy of a forgotten industry. There is a certain spookiness about the place, as if it held secrets better not known. It is one of very few places where I have had the experience of going into a pub and every head turned silently to signal my arrival.

The connection between coal and the industrial city is perhaps too obvious to labour. This is one important reason why the Midland Valley of Scotland became an industrial centre, while the areas to the north and south continued to be sparsely populated. The former prosperity and development of Newcastle-upon-Tyne, and the associated shipbuilding industries on the Tyne, and the steel works at Consett, were predicated upon the coal of the Durham and Northumberland field, just as the great conurbations of Leeds, Bradford and Huddersfield grew up on the Yorks field, and why else did Sheffield become the world leader for the manufacture of fine cutlery? On the other side of the Pennines, the Lancashire coalfield was the basis for the prosperity of Manchester. Great looms shuttled back and forth weaving cotton from the Empire into cloth for export from Liverpool back to other parts of the Empire; mills and mining are at the centre of the dense web of towns to the north of Manchester: Rochdale, Bolton and Wigan. This is cobbles-and-clogs townscape, sentimentalised by Gracie Fields, and now on the tourist trail for those interested in industrial archaeology.

The South Wales coalfield is dissected by rivers that run from north to south, and everything is orientated this way, roads and railways and factories and the grey terraces set singly or in ranks along the remarkably steep slopes. 'The valleys' is synonymous with the coalfield, a shorthand for Rhondda, Merthyr, Rhymney and Ebbw, and a metaphor for much else besides. The heat to which the coal has been subjected in the west of the coalfield has refined it to the highest quality anthracite, rendering off some of the sulphurous impurities. I remember my puzzlement as a child when we took delivery of 'best Welsh nuts' and there were no hazelnuts or walnuts, but very black and slightly sparkling coal. Steel smelting follows, as it should, at Port Talbot, a massive presence by the sea. But the sea breeze drives the wastes from this huge plant towards the sandstone hills, dead skeletons of trees whiten on the slopes, while many more show dead branches protruding from the crown, as if a stag had become ensnared within. Even gorse – tough, persistent gorse – seems leggy and discouraged.

On the 'concealed' coalfield of Nottinghamshire, the country north and south of Mansfield Woodhouse was no less dependent on coal, although here farms and mines persist cheek by jowl, because the rocks overlying the Coal Measures produce good, rich soil. The development of the Vale of Belvoir may add more pits to this Midlands landscape, digging down below to new riches, as yet hidden, although modern sensibilities will attempt to ensure that the intrusion upon the rural landscape is less blatant. This is D.H. Lawrence's country, who described it with fervour, as a 'hot world silted with coal dust'; Alan Sillitoe pointed out that the longer Lawrence was away from it, the more ugly he painted it, a case of absence making the heart grow full of bile. In *Sons and Lovers* and *The Rainbow* one is aware of the unique mixture of the rural and the industrial, where beauty may obtrude, suddenly, in either. Lawrence was obviously aware of an ambivalence in the pitheads, the mean cottages, and the slag-strewn wastes, which could still seem precious in the eyes of his characters. Later in his life, darkness obscured these moments of light.

The association of coal landscapes with industrial darkness is not without irony, because coal is, after all, a product of light. The carbon in the coal was made from trees and other plants that grew by photosynthesis – that is, the translation of light into substance.

In Coal Measure times the vast, tropical delta – its swamps and banks and rivers – sustained life in a profusion never before seen on land. Insects had taken to the air, and thrived there. Colossal, dragonfly-like insects were nearly as large as seagulls. Amongst the litter, cockroaches, not very different from those that still plague restaurants, scavenged for what they could find. Spiders were already catching insects – but probably not in webs, because they had not yet developed spinnerets to produce silk. Scorpions skulked inside hollow logs. Giant millipedes ambled over the forest floor. The forest canopy included tree ferns like those that still grow in lush and humid parts of New Zealand. There were many seed-ferns of kinds no longer living; these included the species I had confused with the royal fern from streamsides in Scotland. The great forest trees included cordaites and lycopods; the latter still has a humble, creeping, living relative, *Lycopodium*, which you would never imagine was capable of such great things. The impressions of the bark of these trees – particularly the lozenge pattern of *Lepidodendron* – are among the most common of fossil remains in the Coal Measures. The roots of the same tree can be found spreading nearly horizontally through the 'seat earths', which often underlie coal seams. Its leaves have been found in another site again, in the coal shales that *overlie* the coal seams: it is scarcely surprising that piecing together an accurate reconstruction of the whole plant was a task that occupied botanists for some time. The final stage in this jigsaw puzzle is identifying the spores or seeds of the plant, which may be distributed to entirely different sediments.

Land-living vertebrates were now lumbering through swamps on four legs, not very speedily perhaps, but firmly enough. Their gait was sprawling, like that of the salamander. Amphibians predominated, and presumably the forest pools were thick with tadpoles. But some of them were large animals with thick skulls, more crocodile than axolotl, and doubtless with predatory habits. There were as yet no frogs, the most successful living group of amphibians, which did not appear until much later. This will serve as a reminder that the different classes of animals did *not* appear in the theatre of life one by one and in order like the characters in an historical pageant. Evolution continued in both what the schoolboy used to learn as 'lower' and 'higher' orders. To further emphasise the point, Stan Wood has recently discovered the earliest *reptile* in the Coal Measures which line the Clyde. As I write, it has no formal name, but it has been dubbed 'Lizzie'. This is the latest of a number of remarkable finds made by Stan Wood in the Scottish Carboniferous: a shark with a great excrescence on its head covered with what look extraordinarily like teeth; a giant scorpion; a fishmonger's slab's-worth of early crustaceans. Mr Wood started as an amateur collector, and has become the palaeontological equivalent of the professional big-game hunter.

From time to time, the sea flooded in over the coal-swamp, thereby abruptly terminating all terrestrial life. This invasion resulted in thin marine bands within the Coal Measures. As the sea spread over the swamps, marine animals began to occupy areas which had shortly before been the demesne of trees and sluggish amphibians. The rocks of the marine bands accordingly yield ammonite fossils. These are crucial for dating the rocks, and particularly useful for correlating internationally between one basin and another. Marine bands of this kind occur in coal basins everywhere: as well as in Britain they are present in Germany, in the United States, in Poland. This distribution has revealed an astonishing fact. The marine bands are everywhere the same age! The only explanation that will fit is that for each marine band there was a sea level rise which translated instantly around the world, flooding low-lying basins, entombing the freshwater and terrestrial deposits that lay beneath. If the same thing happened today, a sea level rise of only a few metres would see Bangladesh and parts of Amazonia drowned immediately. What has further resonance for the present time is that the cause of the sea level rise appears to have been fluctuations in the extent of a Carboniferous ice sheet. There was at that time a polar ice sheet comparable to the ones we have today, and it is not fanciful to draw comparisons between what would happen now if global warming stimulated a partial melting of the ice caps, and what happened in the Carboniferous at the time the Coal Measures were being laid down.

While the same marine band may be traced from one Coal Measure area to another, coal seams changed even from one pit to another; often, the only means of correlating coal-bearing rocks was by means of the fossils of freshwater clams. This inconsistency is matched by a rich vocabulary of local names for seams, not just obvious ones like 'Six foot' or 'Blackband', but idiosyncratic terms, now not a little mysterious. Here are a few from one pit or another in the Scottish coalfields: Earl David's Parrot, Coxtools, Beefie, Humph, Creepie, Kiltongue and Knottyholme. There must have been hundreds of men who knew and worked with these names, cursing them or invoking them by turn. But when these seams are 'mined out' their names exist only in history, retaining a recollected reality only in the memories of miners. I should add that not all coal mining operates deep underground, and that major pits (e.g. in the Fife coalfield) have been worked by the open cast method, especially in conjunction with the exploitation of fireclay from the 'seat earth'.

Coal, or rather coke, came to replace charcoal in the smelting of iron. The marriage of the charcoal industry with the forge had a tradition extending back to the Romans and earlier. Iron produced in The Weald and the Forest of Dean relied on the ancient craft of the charcoal burner. The iron produced by this method was, however, of extremely good quality, which is why so much of it survives; think of the great cathedral doors, or the fittings on offertory chests. The reason for this durability is the very low sulphur content of charcoal iron. Attempts to substitute coal for charcoal produced poor, sulphurous iron – the unwelcome guest from the Carboniferous swamp in another role. However, in 1709, Abraham Darby, working in Coalbrookdale on the small Shropshire coalfield, produced low sulphur coke from coal by kilning it in open heaps in the same fashion as charcoal burners. This cheap and plentiful coke was then adapted for use in the blast furnace, and the steel age was born. It was not many decades before girders transformed engineering, and cast iron added elegance to Georgian squares. These technological achievements are celebrated in the Ironbridge (1777-81) which crosses the upper reaches of the Severn (Ironbridge is now also an outstanding museum). The apotheosis of cast iron is the suitably named Coalbrookdale Gate on the south side of Hyde Park, not far from the Albert Memorial, which somehow manages the contradictory task of combining elegance with monumentality.

The Staffordshire Potteries are famous through the novels of Arnold Bennett: his Five Towns are Tunstall, Burslem, Hanley, Stoke and Longton. The potteries began in a comparatively modest way, but by 1600 the term 'master potter' appears in local records. The clay rocks within the Coal Measures yielded the raw materials for the industry; the Etruria Marl was noted for building bricks, and red or blue tiles for flooring, often known as 'quarry tiles' – a curious misnomer as they are not quarried. The original pots were often big, cylindrical vessels which held 14 lbs of butter, and which were filled and sent by packhorse from Uttoxeter to London. These were common pots of practical intent. But in the third quarter of the eighteenth century, the fine china industry prospered, encouraged and financed by Josiah Wedgwood, whose surname was to adorn

The Ironbridge, Shropshire, symbol of the marriage of coal and iron.

its most characteristic products. Wedgwood imported fine quality Cornish China Clay. This was not an easy operation, given the state of the roads at the time. He sponsored the construction of canals which allowed for bulk transport at low cost. He built a factory, and a village – sometimes called the first garden village – for his work force, and a mansion for his family. He called the estate Etruria: the new canal ran alongside it. The whole operation was the most optimistic and enlightened kind of capitalism, and one might read something of its virtues from the exquisite pottery that was produced there. It was another Josiah, J. Spode, who developed the techniques for making bone-china at the close of the eighteenth century, from which the heyday of Staffordshire pottery grew; by 1800 there were about 150 potteries in the district. Coal subsequently powered the steam engines that crushed and mixed the ingredients for the china, and drove the potter's wheel. The kilns remain to remind us of the great days.

Where there is coal there are sandstone moors nearby. Between the Carboniferous Limestone and the Coal Measures of northern England there are resistant grits of such obduracy that they almost invariably make high ground. These rocks have been known as the Millstone Grit since 1822, and although modern geological usage tends to divide them into several formations the traditional name better conveys character, which is our main concern. The name evokes the time when millstones were an important item of manufacture. They needed to be heavy enough to pulverise the grain, but tough enough not to give way under pressure; the Millstone Grit had the right qualities and locally, there were many water powered mills at the foot of the dales. It is a sandstone cemented together by quartz to form a rock that is often rather pebbly. I visited an abandoned quarry at Millstone Edge, near Sheffield, which was a graveyard of millstones in all stages of completion, tumbling down the hillside, overgrown with bracken. It reminded me of the deserted quarries on Easter Island, where huge, enigmatic carved heads unique to the remote volcanic island lie about in every stage of carving. It is as if the artisans were overwhelmed by a catastrophe so complete that it froze the moment forever.

The Millstone Grit is the record of the passage of a delta which built out over the Carboniferous Limestone, and thereby signalled the end of clear, tropical seas. It underlies country which is best summarised by the word 'bleak'.

The Millstone Grit runs down the east side of the Pennines; patches of it still lie on top of the Carboniferous Limestone to form high features, more than seven hundred feet above sea level at High Seat or Rogan's Seat. It forms the northern, desolate part of the Bowland Forest; on the other side of the Carboniferous Limestone outcrop, north of Leeds, Nidderdale runs its course through it, and Knaresborough Forest is underlain by it, as is Ilkley Moor. Southwards, it forms the southern segment of the Pennines, a central, axial wilderness ten miles or more across separating the two great industrial areas of the North, climaxing in the High Peak, Bleaklow (well named) and Kinder Scout, where the famous mass trespass of 1932 took place.

Haworth was where the Brontë sisters spent their isolated childhood, living in the parsonage, and wandering the Gritstone moors. A sense of landscape is implicit in much of their writings, and absolutely explicit in *Wuthering Heights*, the name derived from the house of High Withens on the bare moorside. Emily Brontë left her wild home infrequently. She must have loved the heather clad hillsides, the mean grass, the dark stone walls, and rushes clustered around boggy rivulets and peaty pools. When Catherine says 'My love for Heathcliff resembles the eternal rocks beneath – a source of little visible delight, but necessary' the

The Millstone Grit: unused millstones lie abandoned by their makers, Chatsworth Grit at Stanage Edge, Yorks.

rocks beneath are the Millstone Grit. Even the name Heathcliff is a kind of distillation of the essential elements of the landscape.

Whether you regard the Millstone Grit as a source of little visible delight depends partly on your prejudices about its virtues as a building stone. It is a tough stone, certainly, but darkens depressingly in the polluted atmosphere adjacent to coal regions. It can go almost black. Where it has escaped the effects of pollution its brown courses can be attractive, especially in conjunction with sandstone roofs. The size of the blocks varies widely according to local conditions; in parts of North Yorkshire and Lancashire large blocks seem out of proportion in small cottages. But local Millstone Grit also served to construct the very grand house at Chatsworth, north of Matlock, seat of the Duke of Devonshire. In the southern Pennines, north of Wensleydale, Gritstone stone walls alternate with those of Carboniferous Limestone, occasionally mixing, precisely reflecting the dominant formation in the local geology.

York Stone is a famous building and paving-stone which derives from the Coal Measures, rather than from the Millstone Grit. Quarries in this formation south of Leeds and Bradford are still working. The flagstones were deposited under fast currents which swept the bedding planes flat, rather than building ripples. For this reason the rock naturally splits into flat slabs. These make the most attractive, ochraceous paving-stones, which can be found all over central London and the West End. Around the Albert Hall there are pavements of York Stone, so, if you are prepared to go down on your hands and knees in public, you may examine the features of the bedding planes for yourself. How much more satisfying such pavements are than those made from concrete slabs, which more and more often have come to replace them. Each stone is subtly different in tone, yet all the differences are a matter of delicate pastel shades. There is nothing more ghastly than seeing one of these stones taken out, in order to put in a parking meter perhaps, and the hole filled in with black tarmac; it is as obtrusive as a boil on a faultless complexion.

The High Peak is an extensive plateau of Millstone Grit sitting on a peculiarly massive bed of cross-bedded sandstone. A scarp edge made of these grits can be examined at Kinder Scout, but the plateau above seven hundred metres is a feature-less place. If a little mist obscures the horizon it can be disorientating, a Sargasso Sea of bilberry, crowberry and heather. It is not difficult to get lost, especially if you leave the path and follow one of the many tempting gulleys that start purposefully only to peter out enigmatically among the heather. These rocky gulleys are peculiar to Gritstone moors, and are known as cloughs. The springy, ericaceous *maquis* can appear almost black at times of the year when the heather is not in flower, and it is impossible not to contrast this dour vegetation with that of the brighter limestone country. The Millstone Grit has a particular affinity for darkness. The plangent cry of the curlew carried across the plain only adds to the feeling that if there were a place where unhappy souls might wander this would be it.

The flora of the acid moorland on the Millstone Grit has much more in common with the moorland of Scotland beyond, and not only the same species of the heather family. There are patches of cotton grass moor which can be matched plant-by-plant north of the Highland Boundary Fault. Sundews (*Droseras*) and Butterwort (*Pinguicula*) and other specialist plants adapted to bogs with little nutrition grow where water fails to drain away. The bog asphodel (*Narthecium ossifragum*) grows here, too, a surprising bulb carrying yellow flowers in July and August. In the seventeenth century, Lancashire women used it as a hair dye. The species name means 'bone breaker' and this probably refers to the propensity of this plant to give sheep the bone disease cruppany. There are all manner of sedges and rushes. Away from the heather moorland, and particularly on more shaly rocks, mat grass covers damp hillsides where sheep impose their braided tracks, making a dull sward relieved by the little yellow stars of flowering tormentil (*Potentilla*) and violet-blue puffs of devil's bit scabious (*Succisa pratensis*). In the Bowland Forest (Lancashire) the Millstone Grit makes wild and inaccessible fells, where grouse are the cash crop, but Clougha Pike, near Lancaster, is more easily approachable; the summer visitor will see ring ouzels and wheatears on Windy Clough, and there is a good flora. In Derbyshire, it is easy to travel from acid moorland flora on the Gritstone to an utterly different limestone flora in the 'pavement' country in the compass of a few hours; there is nowhere better to see the influence of the rock foundation upon the details of even the humblest herbs.

If Kinder Scout or Combs Moss can seem implacable in their openness, the Grit is also capable of weathering spectacularly. Alport Castle (Derbyshire) is on an isolated crag surrounded by slopes of coarse moor-grass. The whole can look like a decayed Mayan temple in the afternoon, when low incident light brings out ruddy tones in the massive sandstones. The Doubler Stones two miles south of Addingham (West Yorks) are isolated weathered stacks of Kinderscout Sandstone which look like tables dedicated to a forgotten cult of sacrifice. Some eight miles east of Ripon (North Yorks) the Brimham Rocks are a fantastically weathered Grit plateau at a height of more than 300 metres. The largest rock pillars are six metres high, terminal monuments to wind erosion, the fretted relics of resistant beds – a miniature, but not quite domesticated version of the semi-desert landscapes of the Wild West. The sculpted stones suggest all kinds of grotesqueries, and the devil is first among them (as in Devil's Anvil), but there are Druidstones as well, and a dancing bear; but it is as much a pleasure to let your imagination play over the stones and suggest new names of your own: is this the profile of a witch, is that a tumbled coffin? Decide for yourself.

There is another great area of Carboniferous grits. This is the central area of Devon enclosed between the Devonian of Exmoor to the north and the vast Dartmoor granite with its surrounding black Devonian slates to the south. This area lay to the south of the clear Carboniferous Limestone sea, and, later, south of the Coal Measure delta. It was a sink for debris from the north, a turbid sea for much of the time, where a few ammonites swam. It is not always easy to find fossils in the sandstones and shales. Devonian and Carboniferous were both folded in the Hercynian movements, and eroded deeply, before the New Red Sandstone

The harder beds of grit cap stacks composed of softer sandstones which are more easily eroded; the Doubler Stones, near Addingham, Yorkshire.

covered both to the east under one of the great unconformities in our geological history. It is not surprising that Carboniferous grits and shales are often steeply, even vertically dipping, and sometimes conspicuously folded. The structure of this area has proved very complicated, and many of the details have only been worked out over the last few years. The wide 'syncline' of the western peninsula is a jejune oversimplification. It has been shown that south-western England includes a whole series of nappes, comparable, if anything, to the Caledonian structures north of the Highland Boundary Fault (Chapter 4). This was, after all, another mountain range, which scraped Britain at its western edge after the Hercynian ocean to south and east was destroyed.

The Devon Carboniferous rocks are known as the Culm. A few rather scrappy and impure coals occur at certain levels, but for the most part these were rocks that accumulated under the sea, even at considerable depth. The valleys of the rivers Taw and Torridge amble through the centre of Culm country. It is a continuously rolling landscape where no road seems to be able to keep its mind on

The southern aspect of the Carboniferous: slates and sandstones spectacularly pleated by Hercynian folding. Near Pennally Hill, Boscastle Harbour, Cornwall.

where it is going. Cottages are mostly rather humble constructions of sandstone rubble. The Culm only comes into its own when it reaches the coast.

Hartland Point lies at the apex of a triangular area of land which projects into the Bristol Channel. It lies about midway along a coastal path that runs from Westward Ho! – probably the only town in the world with an exclamation mark as part of its name – to Bude, all of it in Culm country. Those who feel that the West Country is over-populated with tourists soon escape them on this path. It dodges in and out of woodland at the top of spectacular cliffs; the folded Culm rocks make tremendous height, but the only beings with a full appreciation of the geology must be gulls or buzzards, effortlessly cruising on the updrafts out to sea, where they can see what the walker cannot. The precipices are often overgrown with trees that hang on to the dark cliffs with marvellous determination. Occasionally, there are streams that tumble out to sea along steep-sided little valleys, and then it will be a scramble to gain access to the beach. However, beach is a misnomer. Instead, there are usually jumbled slabs of grit, or rounded boulders of grey sandstone. The Channel waters often have a kind of greyish opacity, even as they suck remorselessly at the dipping beds, seeking out weaker shales between the grits. It is a lowering coast. Grey seals favour it, secure in its remoteness, and their colour seems to suit it.

Men have snatched improbable harbour on this wild coast, building out into the sea to take advantage of the smallest bays. Clovelly is one of these. It is an improbably steep village, ranged down the cliff itself. Cars must park at the top of the hill, so that one appreciates what life has been like there for hundreds of years: hard on the legs.

Hartland Point, by the lighthouse, is a place to look northwards, and imagine yourself back in time, floating upon the Carboniferous ocean; beyond the horizon coal swamps, lakes and streams stretch away far to the north and to the east. But soon their humid plenty will come to an end. Nothing in geology is more certain than change: even the cliffs on which you stand are doomed.

11

Lost in the Sands

THERE IS one kind of Italian opera in which a tenor sings a recitative at a spanking pace to explain all the action that has been happening offstage, so that business can proceed rapidly to the next aria. I feel sympathy with that tenor when it comes to describing the Permian and Triassic. In Britain, the two systems merge together. As far as the history of life in our islands is concerned, it fades during the Permian, and resumes dramatically by the Jurassic; H.H. Swinnerton long ago likened these times to a kind of entr'acte, a curtain drawn across the stage. Before, the Carboniferous world was one of amphibia and lycopod forests, and there were still trilobites in the sea; after, there were dinosaurs at their prime, and the trilobites had gone forever, along with most of the animals which had lived alongside them in the Palaeozoic. The end of the Permian was a great crisis in the history of life, and we see nothing of it in Britain. To understand this part of the hidden landscape, we have to see our islands as part of the planetary whole.

So, taking a deep breath like the tenor about to begin his recital, I will attempt to describe the rest of the world in short order, the better to understand our British corner: for example, why Cheshire was the centre of a medieval salt industry, or why timber frame building reached perfection in the middle of England.

Most importantly, there was the fact of Pangaea. In the Permian, all the continents were fused together as a 'supercontinent': the Hercynian earth move-ments had married the southern continents with the northern. Pangaea, the most fundamental historical fact about our planet in the last two hundred and fifty million years, was the conceptual battleground between those who believed in a fixed and immutable geography, and the continental 'drifters'. For many years

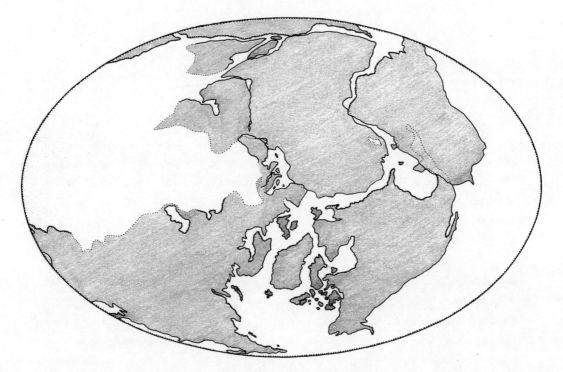

Pangaea, the 'supercontinent' of which the British Isles were but a small part.

it was the latter who were in the minority. Nowadays, it would be difficult to find any scientist who is *not* a 'drifter'. As I described near the beginning of this book, the continents have split and recombined more than once, driven by the internal motor of the Earth itself, patterns which the theory of plate tectonics has done much to elucidate. At Cambridge in the Sixties Sir Edward Bullard, the grand old man of Earth physics, was using computers to generate more accurate maps of Pangaea, an endeavour which seemed extraordinarily sophisticated in those days. New maps seemed to appear almost weekly, but each was a little better than the one before. This was a time when the Earth was being re-invented.

It is now known that Pangaea existed as a complete entity for a short while only, before it once more began to split apart, as the rudiments of the modern Atlantic and Pacific Oceans were created. Nonetheless, there was a time when the terrestrial world was truly as one. In the Permian, oceanic barriers were simply not there, and animals could move until deterred by a change of climate, or, possibly, a mountain barrier. As to the seas, they mostly clung to the edges of the supercontinent. There was not then anything like the

161

length of continental shelf seas that we have at the present day. Shallow seaways sitting upon Pangaea itself were vulnerable to becoming cut off, and, under the glare of the sun, evaporating to sterile dryness.

Pangaea altered weather patterns for the world. Deserts spread. As seas dried out their salt was thrown down in white, crystalline spreads that must have stretched from horizon to horizon. Nothing living could endure this aridity. It is as well that *everywhere* was not like this, or the experiment with living on land that had begun in the Silurian would have come to an end. There would have been no birds, no flowers, no mankind, and only fossils would have testified to a green and various Earth.

Some reptiles became specialised for arid conditions: many reptiles continue to live this way today, because their cold blood revels in heat. While reptiles dominated the larger scales of the ecology, this was also the time when mammals – small, insect-eating, scuttling mammals – took possession of the less conspicuous habitats. And then, when Pangaea split apart, mammals were carried away like rats on a raft: marsupials to Australia, where they stayed and changed, while the rest of the world went a dozen different ways.

Finally, the marine realm suffered its greatest extinction at the end of the Permian, and, as recent research is making clear, was hit again within the Triassic. Everything living in the sea was affected. One solitary ammonite species survived, to give rise to all those of the Mesozoic. The trilobites did not survive at all; nor did any of the corals which had formed the reefs of the Carboniferous. *All changed, changed utterly . . .*

The small piece of Pangaea that was to become the British Isles was in no specially privileged position. Marooned within the huge continent, its history through this critical period was dominated by the refractory geography of the world, and forged in heat, implacable tropical heat. The uplands thrown up by the Hercynian movements were being wasted, in just the same way as, long before, the Caledonian heights had been reduced and transmuted into Old Red Sandstone. Red sandstones and siltstones spread out like aprons around the uplands, filling basins, blanketing older topography. These Permian to Triassic red rocks have been known for a long time as the New Red Sandstone, and they appear as such on all the old maps. Modern texts, like the *Geology of England and Wales* published by the Geological Society of London (1992), have eschewed this name in favour of local formation names, which is a level of refinement that we do not need here. The red sands and silts are dominated by evidence of aridity, something like the sedimentation which is happening now in the wadis of the Middle East; the deposits of flash floods, temporary lakes, migrating sand dunes. The 'New Red' often has a bright – even garish – redness, with a raw edge to it that distinguishes it from the more mellow tones of the 'Old Red'. The colour, which is ferric iron, passes into the soil in the Midlands and Devon, giving it an almost luminous redness when it is freshly ploughed.

A glance at the geological map will show how the Permian marches across the great syncline of Devon and Cornwall, burying its eastern margin; Triassic rocks surround the Quantock Hills. The same rocks flank the mouth of the Severn, on the southern side of which the Carboniferous Limestone of the Mendip Hills and Bristol Gorge stand up like monuments. In truth, the topography now is probably not very different from what it was in the Triassic, but at that time the hills would have stood out bare and gaunt under the unblinking sun, and where cows now graze rich fields, there would have been sand wastes or playa lakes crusted with white crystals and as hot as the fires of Hell. The Severn Valley traces the Triassic outcrop northwards into the Midlands – in fact, one might say that the Triassic *is* the Midlands: Warwickshire, Leicestershire, Birmingham itself, Staffordshire, are all substantially underlain by red rocks. The older rocks of Charnwood Forest (Leics), and the Leicester and Coventry coalfields, emerge today from their Triassic cover, forming hills again as they did when they contributed their waste to the red plains. East and west of Derby, the Triassic blanket buries the south end of the Derbyshire coalfield, and here it splits into two great arms: the western arm encompasses the Cheshire Basin; the eastern arm is twice as long, a great swathe running north from Nottingham to York, and finally reaching the sea at the Tees estuary. These areas were basins when they received their floods of sediments two hundred million years ago, and basins they remain. Rivers seek out the comparatively soft sediments; the course of the Trent follows the outcrop of the Triassic with precision, and there is much of the Severn, the Stour, the Arrow, the Blithe and many more. The river valley silts and weathering of the marly Triassic combine to make one of the most fertile soils in the British Isles, equally suitable for arable or dairy farming.

North again, there is another Permian-Triassic basin centred on the Solway Firth and Carlisle, and continued southwards west of Cross Fell to Penrith and Appleby. Cross-bedded sandstones which are much in evidence around the town centre of Penrith itself are true fossil sand dunes, with every sand grain polished to a perfect sphere – 'millet seed grains' is how textbooks tend

Creswell Crags, Derbyshire. Fissures in the Magnesian Limestone (Permian) are wide enough in places to have provided caves in which palaeolithic men found shelter.

to describe them, thus making a charming assumption that the average student will be familiar with millet seed. Patches of 'New Red' in the Midland Valley of Scotland extend westwards on to the Isle of Arran.

Because they make vales, the areas floored by the Permian and Triassic are natural corridors for traffic and trade. Canals were driven through them, like the Trent and Mersey or Shropshire Union. Their towns – York, Newark, Nottingham – are ancient ones. Nottingham Castle is built on crags of sandstone which command the low ground in the valley of the River Trent. There is an inn hollowed out of the crags, the Trip to Jerusalem, which has claims to be the oldest inn in the British Isles. It is still serving ale under ceilings which seem to merge into high crevices. It must be the only place where you can investigate the peculiarities of sandstones while ordering a glass of sherry.

But the rich soil which the marly rocks produce when they are deeply weathered explains why it has been cultivated for so long. F.W. Maitland distinguished two categories of English countryside, which Oliver Rackham has termed Ancient Countryside and Planned Countryside. The former is thought to have been cultivated more or less continuously for more than a thousand years. Ancient Countryside is the England of ancient grass tracks bounded by hedges on either side, and sunken into holloways by the endless passage of carts and cattle. This is where hamlets with manor and farm and church and pond are tucked into protected valleys, where a maze of public footpaths criss-cross in arcane patterns, where heaths survive, and where the roads twist and turn apparently out of sheer perversity. This is where dozens of small and ancient woods still linger on, which were regularly pollarded and coppiced, and now betray their history only through neglect. Hints of even more ancient habitation – standing stones, ancient fords, moats – give mysterious signs of a history so remote it cannot be recovered, only guessed at.

Planned countryside has fewer, but larger villages. Hedges are straighter, younger (hawthorn predominates), and define larger fields, a 'drawing board landscape' conceived under the influence of the Enclosure Acts in the eighteenth and nineteenth centuries. There are wider vistas, fewer surprises, or ponds, or ancient rights of way, and little coppicing. The roads tend to go from A to B.

The western belt* of Ancient Countryside follows the Permian and Triassic outcrop quite precisely, from Tees to Exe, even to the extent of showing bifurcation into the Cheshire Basin in the west, and the Vale of York east of the Millstone Grit. Prosperity created by rich farming presumably encouraged towns both for exchange of goods and for protection of the people, with the sometimes dubious help of the local baron. This is evidence that the hidden landscape controlled the development of agriculture and its landscape over a period even longer than we might imagine. By and large, the 'New Red' has been a civilising influence.

The prime area of 'black and white' building in Britain includes much of the countryside in the west Midlands underlain by 'New Red' – as well as the adjacent areas to the west occupied by Old Red Sandstone – a country stretching from Hereford and Worcester to Cheshire and including much of Warwickshire and Staffordshire besides. This is where timber-framed buildings peep out at you from every grove of trees, and where they comprise the proudest buildings in the city centres. The woodwork in these western manors and inns was not intended to skulk behind plaster, but to be flaunted, an ostentatious and exuberant display

---

*The eastern belt is broadly the South-east – London and Hampshire Basins plus the Weald – and is separated from the western belt by a wide tract of Planned Countryside including Salisbury Plain and the Jurassic highlands, which became important sheep areas in medieval times and later.

of wealth. Gables lean out, ornamented with quatrefoils, arches and criss-cross designs, dark beams contrasting with white panels, as pied as liquorice allsorts. The most flamboyant houses are mostly late Tudor or Elizabethan. Some examples seem almost *too* chequered, as if display approached parody, like Little Moreton Hall in Cheshire. There is also a suggestion that the chequerboard appearance was a nineteenth-century embellishment – it may be preferable to let oak weather to a pearly grey and avoid blackening. They were built to last from English heart oak, and last they have. The sandstone country, with its covering of alluvium, and glacial clays, evidently suited oaks, and the affluence of the area allowed them to grow to glorious maturity.

Rich dairy country has another virtue: cheese. Cheddar cheese was a product of the pleasant pastures along the Axe valley, not from the spare Carboniferous uplands of Cheddar Gorge itself. Definitive cheddar is still made there, nutty to the taste, and with a white rind as if it had been encrusted with frost. Ersatz Cheddar has gone around the world in the guise of 'mousetrap' cheese, a species of solid block with the texture of soft wax. Cheshire cheese, at its best white and crumbly, is still largely made on the Triassic farms of that county. Red Leicester

Little Moreton Hall, Cheshire, the apotheosis of half timbering built for show, typical of areas where oak was preferred to stone for construction.

has its admirers, also. Even the king of British cheeses, Stilton, although apparently named from a village firmly on the Oxford Clay (Jurassic) of Cambridgeshire, may have only been *traded* there, because Stilton lies on the main route north, these days the A1, which passes into Permian and Triassic territory northwards. Certainly, the best Stilton I have ever tasted was from a Nottinghamshire cheese maker. The excellence of the cheese is some compensation for the paucity of the fossils.

It is not true that the Permian completely lacks fossils. Along the eastern arm of the Permian and Triassic outcrop there is a long strip of ground which is coloured dark blue on the geological map. It extends from Nottingham to Sunderland, with its widest outcrop at the coast north of Hartlepool. This is the Magnesian Limestone. It is so called because it is made from the mineral *dolomite*, which differs from calcium carbonate – limestone – in having the magnesium atom as well as calcium combined with the carbonate molecule. It will come as no surprise to learn that the Dolomites are largely made of it. It is an attractively creamy-yellow building stone. The Magnesian Limestone marks a brief incursion of the sea into our barren islands at this time. This sea extended eastwards to Germany and beyond, and has been called the Zechstein Sea. At about the same time another, saline arm of the sea crept in along what is now the Irish Sea and flooded the Cheshire Basin, lapping against the western edge of the Pennine ridge.

In the Zechstein Sea there were clams and brachiopods and lots of species of sea-mats (bryozoans), occasionally fish, all of which are now found as fossils in the Magnesian Limestone. Environmental conditions were likely to have been similar to those in the Persian Gulf today. One of these fossils is called *Horridonia horrida*, from which you might reasonably conclude that it was a warty, lumpy thing covered with excrescences. In fact, it is a rather ordinary brachiopod with long spines growing out of the shell. Its time did not last long. When connection to the open ocean was reduced, the sea evaporated under the blaze of the tropical sun, to become more and more charged with mineral salts; few animals could tolerate such hot, salty sea. Eventually, brines started crystallising. These natural crystal deposits thrown down from sea water are known as *evaporites*.

When the sea was rejuvenated by a new influx of salt water, the process could begin all over again. Over millennia, thick deposits of natural salts were built up. There are deposits of this kind both to the east and west of the Pennines, in Yorkshire and Durham, and the Cheshire Basin.

These deposits do not, as perhaps might be anticipated, comprise just rock salt (halite). Salt is usually thrown out of solution at the end of the evaporation process. At an earlier stage, the mineral anhydrite ($CaSO_4$), calcium sulphate, crystallises out. This mineral may be more familiar in its hydrated form, gypsum, with which it often occurs. When perfectly formed, rock salt produces exactly cubic crystals; it must be the only rock which can be identified easily by taste. Anhydrite is usually more shapeless in form.

This chemistry is not mere anecdote, because we have to thank the slow evaporation of the sea under the merciless Permian and Triassic sun for the growth of dozens of industries. The exploitation of salt in the Cheshire Basin and near Droitwich has been happening since the Dark Ages. Salt was of immense importance at a time when salting down was a customary method for preserving food through the winter. It is not to be wondered that ancient towns still have their 'Saltgates' (the Norse word 'gate' means street) and 'Saltways'. Now, brine is pumped from its strata underground in the Cheshire Basin, which has caused much subsidence over the Cheshire Plain, and initiated dozens of ponds. The use of salt as a condiment still continues, but salt also lies at the root of many industrial processes. It might be said that as geology is to landscape so salt is to chemistry. The ready availability of limestone in the same area led to the development of the soda industry; this in turn linked with glass manufacture. Hydrochloric and other acids were derived from salt; sodium hydroxide and other alkalis fed onwards into a soap industry: one chemical connected with the next in a kind of commercial chain-reaction. Anhydrite and gypsum are at the base of another chain, of which the several uses for plaster and its derivatives in the building industry are only the most obvious. The mighty Imperial Chemical Industries, founded in 1926 by an amalgamation of smaller companies to compete with potentially even mightier German competition, was rooted, in deep geological truth, upon the heat of Pangaea. The site of its factories at Billingham near the northern end of the Permian-Triassic outcrop still acknowledges this truth.

There is another face to Permian geology, an implacable, hard, unflinching one: granite. Dartmoor, Bodmin Moor, Land's End itself, are all parts of an enormous granite body (batholith). They rise to the surface as would isolated peaks from a massive iceberg, or the humps of some Leviathan visiting from the depths of the Atlantic. Much more remains below, far below, in a landscape that is visible only to modern geophysical instruments. A glance at the geological map will show how Dartmoor cuts through the Devonian and Carboniferous country

A granite tor on Bodmin Moor, the visible tip of the great batholith which extends from Bodmin to Dartmoor.

rocks, and is therefore obviously younger than either. The complex Hercynian movements passed their acme, and the granite was intruded into the mountain range, its light magma rising through the crust, baking it – or engulfing it – as it went. Perhaps a metaphor of a great beast rising from the depths is not altogether inappropriate. Granites were intruded at depth, and their appearance at the surface is a measure of how much of the rock that once covered them has been stripped away by erosion. It is known that the Cornish granites had already reached the surface by the Cretaceous, because some of the distinctive minerals they contain have been found in Cretaceous sediments.

There have been plenty of other granites mentioned in this book, the Caledonian granites of the Cairngorms or Connemara greatest among them. These intrusions have an almost terrifying monumentality in the face of assault by ice and weather; Pleistocene ice sheets grew from them, and even now they look as if the ice had left only the day before yesterday. They are magnificent, but scarcely endearing, recalling Mont Blanc in the Alps. These granites have no bedding planes to yield to the vice of frost; ice carves them by brute force; they are the personification of the laws of scour.

The western granites are different. They lay beyond the margin of the massive ice advance belonging to the Anglian glaciation; they have had time to respond to the elements at comparative leisure rather than being bullied and chivvied by ice into grandiose and fretted peaks. The face of the country is more open, and rolling rather than precipitous, although many a frightened walker, trapped or lost, has had cause to remember that there is nothing domestic about the scenery. Moorland here is just as authentic as Scottish moorland, and there is blanket bog south of High Willhays every bit as treacherous. In the bogs, cotton grass like miniature candy floss nods in the breeze, and heath, heather and gorse take the high ground. There are sheep, of course, but also dark and hardy Dartmoor ponies. Even the birds respond to the imperatives of the landscape, for there are ravens and buzzards and ring ouzels. Dartmoor is a close neighbour of the Highlands in all but miles.

Millennia of weathering have left residual granite peaks standing as lonely tors above bracken and short-cropped grass. Yes Tor rises to more than seven hundred metres. Occasionally, weathering has left strange monoliths. Bowerman's Nose, near Manaton (Devon) is a sentinel composed of blocks that seem to have been piled up almost carelessly. Now it looks out over ancient fields like some forgotten guardian from the time when standing stones were erected in pale imitation of what Nature can do in league with unhurried time. Tors and monoliths alike are what has been left behind by the elements. Because granite is a coarse igneous rock – felspar, quartz and mica stirred together without grain or weakness – the only controls on its weathering are joints, cracks which divide its mass into cubes or rectangular blocks. The largest, or toughest blocks survive to form the tors, then perhaps to have their edges rounded, and their sheerer faces dappled with lichens and mosses which thrive in the moist climate of the South-west.

Buildings made of granite are tucked into the valleys of the River Dart, or cluster around the edge of the moor. Boulders or crudely shaped blocks make buildings which seem scarcely grand enough to deserve the immortality their stone guarantees. H.M. Prison alone stands on the moor in grisly monumentality, and who can see it on a stormy night without a frisson of dread? When I have had a hard day cracking unprofitable stones I sometimes reflect ruefully that what I do for a living is what footpads and brigands once did for punishment; furthermore, it was declared inhumane! There are granite relics of days long before the prison, when ancient packhorse routes threaded their ways across the moor. Clapper bridges cross streams at Postbridge, Wallabrook, Dartmeet, with piers and spans built of solid rock; you can almost hear the rattle

Dartmoor Prison: isolated upon the batholith.

of hoofs as the hastening medieval traveller made his way across dreadful and dangerous wastes to the warmth and safety of an inn at Tavistock.

Quoits are burial chambers whose genesis dates back thousands of years; they are the granitic cornubian cromlechs, standing stones huddled together and capped by a simple slab. They cleave to the granites: on the Land's End mass the Lanyon and Mulfra Quoits, and on the Bodmin boss the extraordinary Trethevy Quoit, which dominates the landscape with palpable authority. Did the makers have some inkling of the endurance of their work?

The Romans came to the South-west in search of tin, and its ore has been mined, off and on, ever since. The last mine may be about to close. Tin and copper compounds were among the exhalations from the intrusion of the deep granite, squeezed into cracks, fissures, faults and interstices in the country rock, so that now the metalliferous veins worm their way along faults or joints, and must be chased through shafts. Both metals are valuable and rare, and were it not for the successful exploration of the world for cheaper sources the West Country would still be a mining centre. Copper has never ceased to be important since it was first used prior to the Bronze Age. It is one of the few metals that can, rarely, be found 'native' – as pure metal, not bound up with carbonates or sulphides which must be removed by smelting. Its capacity for being beaten into shape made it immediately useful in the manufacture of utensils. When copper was combined with ten per cent of tin to produce bronze it was an altogether tougher alloy and more useful for tools, and weapons. Some historians claim this as one of *the* crucial technological breakthroughs in human history. The Redmore Mine near St Austell has been dated to the Iron Age, and may be older. Now copper is as important as ever because of its excellent electrical conductivity. For fifty years in the last century the West Country was the biggest producer of copper in the world. Mining was centred in the Tamar valley, which follows the Devonian ground between the resistant heights of Dartmoor and Bodmin, in a well-wooded and steep-sided valley. Today, solitary chimneys mark the sites of former mines. It is difficult now to envisage beam engines striving noisily in pump houses belonging to Consolidated Cornish Copper Mines and Devon Great Consuls. Miners' terraced cottages survive, brown grits and slate roofs, built on the back of profit which has moved elsewhere.

The Cassiterides ('tin islands') was allegedly the Roman name for the Scilly Isles, which is where the trail of Cornish tin continues westwards. The mineral cassiterite recalls this ancestry: it is one of the metal's principal ores. Around Camborne and Redruth tin mining is still close to being viable. Mining is invariably the least sentimental of business endeavours; a price shift of a dollar a tonne can throw a prosperous community into penury, and Cornish tin mining has seen several such reversals. When the price was 'up' the famous Camborne School of Mines was founded, which survives despite the fickleness of markets, which must say something about the durable values of teaching in a world dominated by economic flux.

The strangest tip heaps in Britain dominate the skyline near St Austell (Cornwall); creamy-white and conical, they look like volcanoes of flour. They are hills of waste from the china clay industry. When granitic rocks weather deeply they leave a clay residue of the white mineral kaolinite, which is an excellent raw material for ceramics. This industrial wasteland is bizarre enough to have been used for film sets of alien planets. When it rains heavily around St Austell the streams run white as milk.

If this book were predicated upon geography it would finish at Land's End. The tip of Cornwall is no whimpering end either, but a defiant, emphatic kind of ending, a full stop terminally punctuating the long tale of the countryside. If

Tin mines in Cornwall are now mostly picturesque ruins, but once supplied British needs. This is Botallack Mine, circa 1870.

the sea rose it would nibble in at Penzance and St Ives and strand the granite peninsula as an island. Something of the kind can be seen every day as St Michael's Mount becomes abandoned to the tides. It is an accident of geology that Land's End is almost the only place in Britain where a great granite boss directly confronts the sea. Sheerness is the rule; mighty breakers seek to probe weaknesses that are not to be found in granites as they are in less noble rocks. A paler zone along the vertical cliffs shows to what height waves are capable of hurling themselves in their winter fury. Who can walk north-east from Cape Cornwall without their feelings wavering between awe and nervousness? Where streams come to sea there will be narrow coves, as at Lamorna, or harbours may be built out improbably to snatch safe haven from the jaws of the Atlantic Ocean (Mousehole). The place to reflect upon this elemental drama is the Minack Theatre at Porthcurno. Rowena Cade created a theatre here, high on the granite

Land's End, Cornwall, a coastline of granite. Erosion by the sea has produced caves, arches and clefts in the cliffs known by the local name of zawns. The rock with the arch is Enys Dodman, and above it lies Armed Knight.

cliffs, hollowed out from the hardest rocks in the land, at the end of the land. The cliffs continue vertiginously to fall below, so that you seem to be perched precariously in mid fall, safe so long as you hold on to incorruptible granite, for all that you can see the white flecked, heaving swell, now dipping, now rising and sucking at inaccessible cliffs. Where you sit was once far beneath the Hercynian peaks, and has been exhumed after so many millions of years to defy the worst that winter gales can do. The granite will lose, in the end, and once more a cycle of the Earth will take a quarter turn. But for now, Man can act out his own ephemeral dramas on the stage below, more temporary than the feeblest lichen, less tenacious than stunted heather that crouches in joints deepened by erosion into obstinate granite.

12

Vales and Scarps: the Jurassic

WHEN I WAS a child I was given a copy of Arthur Mee's *Children's Encyclopedia*. It came in several dark blue volumes. My edition must have been published at some time in the 1930s. It was a remarkably *worthy* compendium, predicated on Kiplingesque notions of service and doing the decent thing. Science was not neglected, and it must have been from these pages that I first glimpsed something of the intricate complexity of the world. Spread through the volumes there were short pieces on geological time. Each geological system was illustrated with a diorama of 'life in the so-and-so period'. My favourite was 'Life in the Jurassic period'. Giant sea lizards stiffly poked their noses out of a sea full of ammonites and cumbrous fishes plated in bone. On a neighbouring page there was an account of Mary Anning, who discovered and excavated some of the first, magnificent, giant sea lizards (ichthyosaurs) around Lyme Regis in Dorset. There is a portrait of her in the Natural History Museum, with a basket over her arm, as if she were going to glean a few crocus bulbs to exchange with the rector's wife rather than to expose some of the most elegant and extraordinary of extinct marine creatures.

So Lyme Regis seems to be inextricably linked with palaeontology in the general psyche. The hero of John Fowles' *French Lieutenant's Woman* was after fossils in Lyme when he made an early acquaintance with the woman who fascinated him. It is a delightful town, regardless of palaeontology, and particularly out of season. The sea front is as elegant as can be, and the whole town is ranged in tiers upon the cliff sides with scarcely a discordant building to be found.

But if any ground in Britain deserves the over-used epithet 'classic', it is the Dorset coast, extending from the environs of Lyme eastwards to Swanage. Although the Jurassic System takes its name from the Jura Mountains in southern

France there is probably no better conspectus of the whole caboodle than along the cliffs of Dorset. Starting with the earliest Jurassic rocks in the west, the formations proceed one after another along the coast to Portland Bill; the higher formations are repeated again on the Isle of Purbeck. This is the southern end of the great swathe of Jurassic which runs north-westwards to the Cotswolds and thence northwards to divide the Midlands from East Anglia; and further north, coming finally to the massive sea cliffs between Whitby and Filey, where once again the succession can be followed along the coast. This belt of Jurassic, and much of its detail, was clearly shown on William Smith's pioneering geological map of England (1815). With even a modicum of geological knowledge it is still possible to trace out the map upon the ground, because the Jurassic is drawn in rocks with contrasting endurance to weathering, which make vales and hills by turn. Fossils have played an important part in deciphering the Jurassic story, ever since William Smith showed how they could be used to fingerprint formations with diagnostic accuracy. The British Jurassic has something of everything: dark marine clays and shales; limestones forming noble scarps; freshwater deposits with hints of tropical forests; mammals and dinosaurs. This is a return to a full and prolific record of life after the exigencies of the Triassic. The transition from one to the other was gradual, and is recorded in rocks exposed in cliffs and foreshore along the mouth of the River Severn and in Pinhay Bay, west of Lyme. The return of the sea gradually became a flood. These transitional beds are the Rhaetic, named from their well-known development in Germany; good cliffs in Rhaetic rocks can be examined near Bristol Suspension Bridge.

I must mention one of the rocks that is found in the Rhaetic, because it relates rather directly to the theme of this book. This is the landscape marble, commonly called the Cotham Marble, after the village near Bristol where it is better known than in Dorset. This is a limestone bed which, when polished at right angles to the bedding, shows a delightful 'landscape' of 'trees' as tall as your finger; they are reminiscent of the kind of elm trees one used to see in Suffolk before Dutch Elm Disease turned them all into skeletons. Concealed in this rock there is a hidden landscape within the hidden landscape . . . It might have served as a symbol for this book, but it fails lamentably to be appropriate, because the trees are no miniatures; rather, they are the structures of an alga, a kind of sculpture made by these simple plants within the sediment. The rock also provides an instance of how the term 'marble' has been misused for any limestone which can be polished to ornamental effect. The geologist's marble is a metamorphosed limestone, transformed by heat and pressure so that any trace of organic remains it may once have contained have been obliterated. So the landscape marble provides a fake landscape in a spurious marble – a feast of terminological inexactitude.

'Landscape marble' from Cotham, sculpted by algae to give a picturesque landscape of 'trees'.

The Lias, more particularly the Blue Lias, is the rock formation at the base of the Jurassic. It can be examined in Chippel Bay just to the west of Lyme, or at the other end of the outcrop in Redcar, on the Yorkshire coast. 'Lias' is how the 'layers' were pronounced by the local quarrymen in Dorset, and layers are what it has. Prominent limestone layers form ribs separated by shales, and are easily extracted in slabs or 'bricks'. They have been used to make rather densely-grey houses and walls around Lyme, and in the Polden Hills (Somerset)*. The Lyme quarrymen gave each of their layers special names: Specketty, Mongrel, Rattle . . . names which have only ever been known to a handful of craftsmen, and now conjure an industry and a connection to the landscape that has been lost forever. There are fossils in the limestones, and often they have been broken up and washed into rock pools. This is where boys and girls can go 'fishing' for fossils at low tide. Most likely, they will find a brachiopod, or a bit of a clam, or a fragment of an ammonite. You can see 'ghosts' of ammonites a yard across on fallen lias boulders lying on the shore at Lyme: ammonites are the spiral ramshorns which once housed squid-like animals. They evolved so fast that they provide a chronometer for the Jurassic which extends around the world, and ammonitic savants can tell the age of rocks from their ribbing

---

*Ham Hill Stone is an iron-rich building stone from the top of the Lias, a rich brown stone much admired, and used to construct the grand house at Montacute, as well as giving much of the character to Crewkerne and Ilminster.

This ammonite, Dactylioceras, *from the Lias has been embellished with a carved snake's head, a tribute to the miraculous petrifying powers of St Hilda (Robin Hood's Bay, Yorks).*

or whorl shape or a dozen other subtleties. The monstrous fragments must have come from something like a giant squid – a huge predator.

By now, the Lias sea must have been deep, the limy layers are rarer, and into that deep sea swam ichthyosaurs in search of fish. They were superbly streamlined, and have been compared in their construction to porpoises, a comparison that has some justice, but is not *perfect*. After all, ichthyosaurs have fish-like tails, which porpoises do not. Fossil embryos preserved in the body cavity show that ichthyosaurs bore their young alive into the sea where they no doubt learned independence very quickly – or died. On the other hand, plesiosaurs do not have any easy analogues. They were unquestionably another kind of sea reptile, but not like a porpoise, nor any kind of seal. It is perversely satisfying that nature could concoct a combination in the Jurassic never seen before nor after: a fish-eater longer than a man with a long, often flexible neck, a seal-like body with flippers, but with an extended tail. It clearly worked very well as a design because plesiosaurs lasted as long in the sea as did dinosaurs on land. It is much less speculative that plesiosaurs ate ammonites, because munched ammonites – complete with toothmarks that 'fit' – have been found in the Lias.

Mary Anning sold an almost complete skeleton of *Plesiosaurus* to the Duke of Buckingham for £200. To extract such a skeleton was the labour of many months. She needed the money to support herself – £200 was a considerable sum in the early nineteenth century – but was there ever better value for money? *Plesiosaurus* must have appeared in a thousand books since its laborious extraction, and continues to intrigue children. Plesiosaurs are now so familiar that it is easy to forget that the sleek creatures were accurately reconstructed only after the most prodigious labours on the part of women and men whose names are now far less well-known than the creatures they discovered. There is a kind of modest heroism in this self effacement in the cause of the past: there are even rules to say that you cannot name an animal, scientifically at least, after yourself, thereby preventing a more devious route to immortality.

Thus something is known of life in the muddy Jurassic sea, and the same description will pass muster for many of the other clay intervals within the Jurassic sequence of rocks in Britain. At times the sea floor got quite nasty – so short of oxygen that nothing could prosper. At other times shell beds grew, or there were limy beds rich in the cigar-shaped fossils known as belemnites. Sandstones formed periodically, especially at the top of the Lias. These variations may have been under the control of the depth of the sea, or so recent theories have it. The details of the succession of beds in the Lias can be followed westwards along the coast through the broad cliffs and rounded hills that abut the sea: Black Ven, Stonebarrow, Golden Cap, Thorncombe Beacon. Michael House has written a guide to the Dorset coast explaining the geology, bed by bed. It is one of those places where you can go straight from a diagram to a cliff, and see the progression of geological time through the early Jurassic, written in the strata as clearly as a first lesson should be, but seldom is.

I should mention that the landslips to either side of Lyme, which make for a delightfully irregular and picturesque walk from Lyme westwards to Pinhay Warren, have nothing to do with the Lias, but everything to do with geology. The Gault Clay is a Cretaceous rock that overlies the Jurassic, and is overlain again by the Chalk. The thick and dense Chalk fractures and slides on the slippery Gault, and it is this that makes such a tumbledown landscape. The permeable Chalk and the impermeable Gault and Lias are juxtaposed in all sorts of ways, which is why there are springs and little ferny gulleys. The Chapel Rock west of Pinhay Bay is a chalk lump that is even now pulling away from its moorings and will one day slide further – who knows how far? This is as confused an area, as the Jurassic coast is explicit.

Further to the east from Lyme Regis, near West Bay, I found one of my first ammonites: I still have it. It is preserved in a limestone the colour of clotted cream. It is a little smaller than a discus, a good, solid object, and wound around upon itself in five coils. It is ribbed and corrugated, to strengthen it. It is not a perfect specimen – part of it is bruised and fading – but it is a precious one for all that. I remember the circumstances of finding it very clearly. The cliffs near

Burton Bradstock and West Bay are composed of the highest unit of the Lias, a sandy formation called the Bridport Sands, which is not rich in the remains of extinct animals. But the cliffs are capped by the next formation above the Lias, a limestone termed the Inferior Oolite. It is a tough rock producing overhangs here and there, which could be dangerous. This is the level from which the ammonite came. The owners of a golf course at the top of the cliff must have become anxious about the overhang and had dredged it upwards, so that blocks of limestone were lying higgledy-piggledy near the clifftop: this is where I spotted my ammonite; it just pulled out. Several years later I found out that it was called *Parkinsonia*. When I revisited the site recently, I did not find any more specimens of the same ammonite, and by then the best place to look at the Inferior Oolite was in fallen blocks, but I did see the famous 'Snuff Box Bed', a nodule bed a few inches thick, so full of iron it weathers to the colour of a rusty nail. It is dotted with laminated concretions, apparently produced by algae, which are the 'snuff boxes', and which they resemble hardly at all. This kind of duff analogy adds to the fun of rootling among the rocks.

The Inferior Oolite might lead to another kind of misconception. 'Inferior' does not mean that there is a rather better quality oolite elsewhere. It refers rather to its position beneath, or inferior to, the Great Oolite, and not to any measure of dissatisfaction. Geological terms are often charmingly anachronistic in this way.

Oolites are composed of perfectly round spheres, usually a millimetre or so across, and very like fish roe. They are not called ools, which is bad luck for Scrabble players, but ooliths; the 'oo' refers to the Greek root for eggs. Ooliths are very easily seen under a hand lens: if you examine any broken surface of a Jurassic freestone you will likely see these perfect egg replicas. Lime ooliths are forming today in shallow, tropical seas in the Bahamas, where particles are kept in a constant state of agitation so that lime (calcium carbonate) can be concentrically deposited around them, like the sugar around the aniseed in a gob-stopper. The arrival of these limestones clearly signalled a profound change from the clays of the Lias: to shallower, clearer and more turbulent seas which favoured shoals of oolitic limestones.

The contrast between soft clay formations and hard limestone formations (plus local sandstones) spells out the landscape – and the character of the older towns and villages – over the broad tract of the Jurassic outcrop. Where limestone predominates there are uplands, or substantial hills, and the local stone has provided stone walling, building materials for the cottages and some of the grander houses, as well as influencing the kind of farming or industry that could be successfully established. Between limestone hills, valleys floored with stiff clays show the other personality of the Jurassic, with heavy agricultural land, and a plentiful supply of local bricks and roofing tiles, which colour the older towns a warm and prosperous red. Thus it can be said that oscillations in the climate and water depth of the sea that flooded England more than one hundred and twenty million years ago still control the uplands and lowlands, the crops and the towns, in short, the whole cast of the landscape.

Because the trend of the outcrop of the Jurassic rocks runs from Dorset to Yorkshire, the belts of the landscape run very broadly in parallel. There are important changes in the rocks northwards into Yorkshire as well, notably an increase in sandstone at the expense of limestone, which imposes a special personality on the North York Moors. In the south there are three clay formations: Lias, Oxford Clay and Kimmeridge Clay, in ascending order, each followed by a limestone interval (plus sandstones here and there), which interpolate between the clays. There are lots of local variations – bits of the sequence missing or thinning or changing character – which do not require description here. The late W.J. Arkell described much of this detail, which is continually being refined, in a massive, 681 page volume on the Jurassic in Britain (1933). This is one of those masterpieces known to few people, but which summarised lifetimes of scholarship and collecting. Arkell was of the Oxfordshire brewing family, and so the barley that provided the family fortune grew upon the Jurassic that Arkell devoted his life to studying. Not content with Britain he went on to describe the Jurassic of the world.

The oolitic limestones lie on top of the Lias and make the Cotswolds, and Bath; they can be traced as hills from Bridport on the Dorset coast virtually to Yorkshire with hardly a break: one of the more blatant impositions of geology upon landscape. Even if it were not easily visible in places, the outcrop could be mapped from dwellings constructed of 'Cotswold stone'. Some cottages are little more than rubble; others are carefully coursed; yet more are faced with worked stone known as ashlar. The oolitic stone is of widely varying quality and workability, and it is curious that some of the stones of lesser quality produce some of the prettiest effects – probably because of their mottled and irregular surface when they are weathered. There are variations in colour from grey to buff, or yellow, grading through ochre to almost red.

Bath itself is probably one of the most harmonious cities anywhere, and the stone has everything to do with this: some find its terraces and parades *too* creamily uniform. The freestone from the Great Oolite has the special property of being easily cut when freshly quarried, so producing flat-faced ashlar of great reliability. In the eighteenth and early nineteenth centuries it spread far beyond Bath and Cheltenham, especially to London; it was in demand, and great quarries operated on both sides of the Somerset-Wiltshire border to satisfy the market. This export was possible because of the expansion of the canals; the Kennet and Avon Canal touched Bath and moved its stone on to other canals and thence around the country. It is a nice paradox that the cutting of those same canals was one of the factors which allowed William Smith to construct his famous geological map showing the outcrop of the Great Oolite. Even today, there are little canal houses along the Kennet and Avon Canal constructed of Bath stone, which make a very odd sight when they are on the chalklands. They look uncomfortable, out of place, slightly embarrassing, as if they had come adrift from the rest of a terrace somewhere and floated off on their own upon the canal. Even more curious is a large house by the Thames which one can see from the train at Maidenhead; clearly, the builder admired the substantial houses of the Cotswolds, but a building of rough stone which might have slipped away unnoticed near Malmesbury, is here miles and miles from its geological home, and serves only as an object lesson in how buildings should spring from the geology beneath them.

The particular feature of a Jurassic limestone house is the roof, which should also grow from the same geological source. Jurassic 'slate' roofs are not slates in the geologist's sense, which are cleaved mudrocks such as the Cambrian slates of Nantlle. They are simply limestones with bedding so thin that they yield flat and durable rectangles for roofing. The wonderful thing about them is that they are graded in size across the roof – largest on the outside at the eaves, progressively smaller towards the ridge – so that no two roofs are identical. This arrangement ensures that reality matches perspective, or even serves to underline it, and the natural colour variation of the limestones guarantees satisfactory mottling. Thin bedded Jurassic limestones have been used in this way since the thirteenth century. With the addition of gables, houses roofed in limestone cannot be bettered; these are the roofs that complete the villages of Gloucestershire, North Oxfordshire and Northamptonshire. They are also very heavy, so that there is always the temptation to replace them with something different. The results are never happy.

Roofs of Collyweston 'slates' − a flaggy limestone − are a most harmonious mix with Jurassic stone walls.

Two famous Jurassic 'slates' are those of Stonesfield, Oxon, and Collyweston, Northamptonshire. Quarrymen referred to the thin-bedded layers as 'the pendle'. The Stonesfield slates are famous for yielding fossils of very delicate vertebrates. Pterosaurs, the elegant flying reptiles of the Jurassic, are known from these slates, and, more remarkably, so are mammals. The Jurassic was arguably the heyday of dinosaurs, when gigantic herbivores and matching carnivores dominated the landscape. But elsewhere, small mammals were flourishing − who knows, they may have actually outnumbered their puffed-up reptilian masters? − but their delicacy militates against their preservation. If shrews were the most numerous species today, it is debatable whether the palaeontologist of the future would realise it. The Stonesfield slates are perhaps the nearest we can get in this country to the lithographic limestones of Bavaria, famed for preserving the early bird *Archaeopteryx* in the sticky mud of a Jurassic lagoon. In this deposit, even the merest pinnules of a feather are preserved. So perfect is the preservation that the astrophysicist Sir Fred Hoyle, in pursuit of a theory, pronounced the fossil feathers fakes, thus impugning the reputations of several generations of eminent palaeontologists. Fortunately, nobody believed him.

The Cotswold villages − small towns might be more appropriate − are not merely bus stops on the Grand Tour of Olde England. They have real merit. They are the apotheosis of Jurassic stone. Towns like Broadway, or Chipping Campden, have hardly any undistinguished buildings, even the humblest cottages are worthwhile. They were built on the backs of sheep, on the prosperity of the wool trade. Stone is pressed into service for everything; no wooden windows; rather, there are mullioned windows divided by stone columns; walls of stone, market crosses of stone, bridges of stone. In Bradford-on-Avon, the oolite fancier's home, there is even a little stone gaol in the middle of a stone bridge in the centre of a stone town. The same town has a Saxon chapel miraculously saved from oblivion because stone buildings are easier to re-use than to demolish. The Cotswold Way has now familiarised walkers with the individual personalities of numerous Cotswold villages. I feel most closely drawn to another wool town, Northleach (Gloucs), because my ancestral roots come from thereabouts. In the charming church there is a famous brass, the Fortey Brass, celebrating the worthiness of one John Fortey in the late Middle Ages, who did philanthropic things locally, and no doubt was the big cheese at the time. We must have been going downhill ever since. Naturally, John Fortey was a wool merchant, and lived in Fortey Hall. The rest of Northleach is another charming wool town. It would be difficult to overestimate the contribution that Jurassic oolite has made to housing the average Englishman in the midwest. Cirencester, Stroud and Malmesbury are further substantial and characterful towns, and there must be hundreds of farms and hamlets built of grey and cream stone.

Around Northleach, and following the trend of the Jurassic outcrop, the Gloucestershire and Oxfordshire plains are wide, flat uplands which directly reflect the nearly horizontal bedding of *resitant* oolitic limestones. You could almost be in Chalk country, were it not for the ubiquity of neat stone walls.

Stone walls are the hedges of Jurassic tablelands. Horizontal courses of rubbly stone 'bricks' of various shapes and sizes are capped by a row of vertical blocks, set at right angles to the lie of the blocks below. At entrances, piers of larger stone blocks serve as secure gate hangings. These walls are tough if well made, and contribute immensely to the variety of the bleaker parts of wheat-belt Gloucestershire and Oxfordshire, as well as offering refuge to lime loving plants. There are signs of a revival of walling techniques, so all is not lost. Barbed wire is barbarous wire by comparison with real walls.

At the western edge of the Gloucester plateau a great scarp along Birdlip Hill marks the edge of the oolites – westward again lies the vale of the Lias upon which Gloucester lies, while on a clear day, you may look much further back in time into Palaeozoic Wales. The great towns below you – Gloucester, Cheltenham, Bristol – do not favour the exigent limestone country. Near Stroud, the rivers cut deep into the limestone plateau, and small stone cottages seem to cling to precipitous slopes as if they were themselves some manifestation of bedding. Near Nailsworth, by fast flowing streams there are large, squarish buildings on several floors. Stone built, solid, permanent, but with large windows in many cases, they seem unlikely heralds of the industrial revolution, but so they are – factories! These early mills used water power from the stream, and wool from the sheep on the hills for miles around – hence, all those Woottons ('wool-towns'), and Woolchester. The wool was cleaned using Fuller's Earth, and this is the deposit underlying the Great Oolite, which comes to the surface exactly here. Fuller's Earth contains a mineral, montmorillonite, which can bind on to grease. These features of the hidden landscape conspire to make this pretty corner of England a cradle of industry.

Near Leckhampton, centuries of weathering have left the Devil's Chimney, like a miniature version of the Old Man of Hoy on the Orkney Islands, standing like a totem pole. Its oolitic limestone has resisted the buffeting of wind and the solvent action of rain. It is reminiscent of a monolith, though it is wholly natural. It is a curious fact that the devil is often named in geological features. It may be that they seem to go against Nature, and therefore must be the work of an evil force. The monolithic stack of the Leckhampton devil suggests the petrifaction of a mythic giant, a Lot's wife of ancient Albion, punished for who-knows-what transgression.

There is a Devil's Chair on the outcrop of the Stiperstones Quartzite (Ordovician, Shropshire), where the weathered relics of tough sandstones perch on the brow of the hill in a crude caricature of a throne. Where better for Old Nick to take his ease as storms track in across the Borderlands? In the Lower Palaeozoic wastes of the Southern Uplands, the Devil's Beef Tub, five miles north of Moffat, was described by Sir Walter Scott (in *Redgauntlet*) 'as if four hills were laying their heads together to shut out daylight from the dark, hollow space between them. A deep, black, blackguardly-looking hole of an abyss it is, and goes down as perpendicular as it can do.' Another Devil's Chimney is on the south coast of the Isle of Wight, near Ventnor, but this time it is a fissure in Cretaceous sandstones formerly cut out by the vigorous scouring of water. It twists and turns, following the natural joints in the rock, like a miniature (but dry) gorge, dark and clammy perhaps – but why invoke the Devil? Then there is the Devil's Punchbowl in the Weald (Hindhead, Surrey) where a great amphitheatre opens out on the junction of Greensand with underlying clays, and here perhaps the devil scooped out a colossal hole to hold his drink; and there are springs along the edge of the Punchbowl, where permeable sandstone joins impermeable clay – enough to supply any satanic horde. Then the Devil's Dyke is an exceptionally steep-sided, dry coombe valley eroded into the chalk scarp near Brighton, a fine example of its kind perhaps, but scarcely diabolical. There is the Devil's Kitchen (Twll Du) in Snowdonia, another chasm, down which tumble waters from the lake of Llyn-y-cwn through Ordovician lavas.

Each of these features is spectacular enough to have invited speculation, and the easy rationalisation is that the devil may be ignorance dressed up as explanation. Before geological time in its immensity had become available as a tool for explanation how else to explain the great gouges, those extraordinary pillars, the steep valleys lacking water? Add some creepiness experienced on a windy, moonlit night, that eerie feeling of being watched, and I would be almost willing to vouch for the devil as natural sculptor, whatever reason tells me. The devil strode widely over the British countryside. There are churches dedicated to St Michael, slayer of devils, particularly in the West Country, and even more particularly churches built on top of hills and tors. St Michael's Mount is well known, but the tiny church of St Michael on top of Brent Tor is equally dramatic. Was the devil resident there before the church? Or perhaps an older, pagan past was exorcised by the dedication? There are old burial mounds in these places. The devil alone seems to match the drama of the sites to which he has donated his name. Strangely, angels do not appear to have inspired many geological tags. Odd rock formations seem to ally with the dark side of the super-natural, and on misty days, sometimes, they still do.

Hell is 'down there' in the hot interior of the Earth, where rock is minted. Is this where the devil must come from? The depths themselves are like the collective unconscious, a hidden landscape that underlies all our natures: feared, probed with reluctance, a commonality shared between all men, which may include archetypes best forgotten.

It is no longer economic to quarry Oxfordshire building stones from many of the local formations which would readily yield them. But there has probably never been more of a premium on the 'character' that such stone imparts. This, coupled with a desire on the part of local authorities to preserve the integrity of their towns, has resulted in the development of imitation stones. These are usually made by cementing together chippings of an appropriate Jurassic limestone from one of the large quarries.

Today, there are neat little estates built of such imitation Jurassic, surrounding the old centres of such towns as Witney or Stow on the Wold. In Witney the supermarket and an industrial estate are built of this material. But, why do these new developments feel less than convincing? On the new estates the neatness, uniformity and economy of the design has something to do with it: houses are tucked into closes and cul-de-sacs, the arrangement of which has much to do with maximising the builder's profit per square metre and nothing to do with producing a passing facsimile of an English village. But the most important flaw is in the bricks. The factory can turn out only standardised sizes, and standardised sizes can be budgeted for and planned for and offer no surprises or awkward corners that might embarrass the bricklayer. The bricks come in two sizes which tend to be laid in alternating courses: this regularity reinforces the impression of sameness in the designs of the houses, even when pains have been taken to ensure that no two houses are the same – or not *exactly* the same. The smooth finish of the bricks ensures that they have something of the sheen of the ashlar finish of a substantial house, but the brick size would be more suitable for a cottage. Pulling between these two traditions, the results are unsatisfying. Conversely, we have the spectacle of a supermarket pretending to be a cottage. Time, as so often, may soften and mottle these buildings, but they do demonstrate that there is more to building in sympathy with the local environment than simply matching the colours.

While warm, shallow Jurassic seas were accumulating limestones, and shoals of ooliths were growing over the whole of southern and western England, there was a very different scene in Yorkshire. In a delta of a huge river, which alternately advanced and retreated from the North, quite different types of sediments were accumulating, which are today seen as massive sandstones and thick shales. In the Cleveland district there are also famous ironstones, which have been exploited commercially in the past (the Frodingham Ironstone in Lincolnshire, which is still worked today in huge open cast pits, is somewhat older – a particular facies of the Lias). The old maps used to refer to these northern Jurassics as 'Estuarine' Series, in prescient recognition of the fact that they were most unlike the rocks in the South, although few of the rocks were really deposited in ancient estuaries. On the coast, they are beautifully exposed from Scalby Ness to Long Nab, and thence around Hundale Poin to Cloughton Wyke, Sycarham Cliff, Beast Cliff, and ultimately Ravenscar. The cliffs get higher and higher along this profile, except where cut by several 'wykes' (steep-sided valleys); until at Ravenscar they are most impressive. There is a waymarked geological trail here, which will take you down the cliff, and down through Jurassic time, until you eventually reach the 'mermaid's tables' on the rocky shore. These are residual sheets of tough 'dogger' which overlie the Liassic rocks and protect them from eroding away. Northwards lies the steep Peak Fault, which allows older, Lias rocks to be seen in Robin Hood's Bay. It is a long walk back to the clifftop. The geological visitor should stay in Whitby, close by, an ancient fishing port set steeply by the sea which could not be more delightful. St Hilda was Abbess of Whitby, and the Lias ammonite *Hildoceras* was named for her. According to legend, ammonites were snakes petrified by the saint. They are extraordinarily abundant in certain parts of the Lias around Whitby, which leads one to wonder whether this can be taken as a measure of the abundance of snakes in medieval times . . . The sandstones forming part of the south cliff at Scarborough were also part of the ancient delta, although ubiquitous boulder clay – the detritus of the last Ice Age – caps the cliff, and often slips to obscure the 'real' rocks. Scarborough ought to be the northern equivalent of Lyme Regis, and it has grand buildings; but it has fallen, as 1930s whodunnits always said of reduced aristocrats, upon 'straitened circumstances'. Out of season it still retains a surprising dignity.

There are dinosaur footprints to be found at some localities. At rather more, there are the remains of plants, some of which, like horsetails (*Equisitum*), lived in the freshwater swamps of the great Jurassic river, but most of which were carried in from elsewhere. The Yorkshire Jurassic rocks are famous among botanists for the knowledge they give us of the flora of the time. You can crack out whole leaves, beautifully preserved, near Whitby. Occasionally, there are thin seams of true coal, and in the sandstones, drifted logs. The late Professor Tom Harris devoted much of his life to making several hundred species of extinct plants known to the world. There are strange things among them, halfway, it has been claimed, between conifers and flowers. But the flora also includes *Gingko*.

This is the maidenhair tree; one member of the genus is still living today. The living species forms a substantial and attractive tree, with unique, thick, fan-shaped leaves a few centimetres across. They turn lemon yellow in the autumn. A fine old specimen is in the Royal Botanic Gardens at Kew. It is often to be found in the gardens of stately homes, having attracted the attention of the botanically curious, who included many aristocrats in the eighteenth and nineteenth centuries. There is an 'espalier' specimen of great size trained against one of the science faculty buildings in the middle of Cambridge. *Gingko* is a lone survivor; there were numerous types of gingko-like plants in the Jurassic, as the fossils gathered from the cliffs and moors of Yorkshire tell us. Why did this one alone survive? Was it merely good fortune, or was it blessed with some enduring quality? Until its recent introduction around the world it lived on only in the gardens of China. The female tree has a fruit but it is no rambutan or mango – it is rather unpleasant. But it does have a place in the Chinese pharmacopia, and maybe it was this property that preserved it from destruction; neither luck nor skill, but an ingredient in a nostrum.

There are also cycads and some of their relatives. They look a little like palms – but they are unrelated to this group of advanced flowering plants. Cycads survive in tropical climates, and quite a few species of them; but they, too, are reduced in variety compared with the Jurassic. I once cut my leg on a cycad leaf in a New South Wales rain forest, and realised that the same sensation might have been felt by an unwary dinosaur more than a hundred million years before. We have to thank a river long vanished, a river that once reached the sea in Jurassic Lincolnshire, for permitting us to imagine the dinosaur's landscape, which might otherwise have been hidden from us; another hidden landscape.

The sandstones which are responsible for the steeper cliffs at the coast also take the high ground on the Cleveland Hills and the North York Moors. The tops of the hills are often capped by one particular massive sandstone known as the Moor Grit. This same rock underlies wild moorland, so dense with heather that it can appear lowering, almost black, until it is transformed into billowing purple when the heather flowers. York Moor country may be the Yorkshire parallel to the Cotswolds in time, but is wilder by far.

The streams draining the moors have eaten deeply down into softer rocks underlying the uncompromising grits, so that the dales and lesser valleys are altogether gentler and more fertile. For the most part, little roads closely follow the rivers, which drain southwards to the River Derwent in the flat Vale of Pickering, the site of a famous Ice Age lake, or northwards into the Esk. It is easier to pass up and down the length of Farndale or Rosedale than to cross between them. Intensely green fields extend up the steep valley sides, following the geology of the less resistant rocks, until they meet the grits of the moorland, where they stop abruptly and give way to open country. Farms are scattered along the dales; nowhere do they brave the moors. Small copses of oak, and rowan, and other broadleaved trees, lead one to regret attempts at coniferous afforestation on the moors themselves. Warm, yellow-grey Jurassic sandstones are used to make the farmhouses, outbuildings and stone walls alike, the latter often rubbly affairs compared with their Gloucestershire counterparts. But there are no limestones suitable for roofing in this area, and red-pantiled roofs are usual. Many of the farms have two stories, with a low byre attached, and a separate barn. Most of them date from the eighteenth century; they avoid the slight feeling of meanness that older buildings in the Dales sometimes convey. The whole scene is delightful – and just about everything about it is rooted in the geology. Because of the exceptional beauty of the area, the North York Moors are now a National Park. It is likely that the connection between rocks and scenery and human use will remain explicit in these charming valleys in the coming decades, and if that is purchased at the expense of a little over-artificial protection, and too many tourist cars, so be it.

The sea returned to Yorkshire in the later Jurassic, drowning the ancient delta beneath a limestone known as the Cornbrash, a distinctive rock which can be traced in one form or another the length of the English Jurassic outcrop: it was mapped by William Smith himself. Thereafter clays and limestone rocks alternate, and it could be said that the history I have related of the Lias and the overlying limestones is repeated twice more, although there are many interesting variations on the more local scale.

It should be mentioned that there are areas of Jurassic rocks far outside the south-eastern belt. Where there has been protection from erosion by a cover of Tertiary volcanic rocks, much of the Jurassic has been preserved in the North-west Scottish Islands: in patches on Skye, Raasay, Mull. They are similar rocks in many ways to those of the Yorkshire coast. Lias rocks also extend into South Wales along the coast near Barry. Also, I remember the surprise when a borehole drilled just near Harlech Castle (at the beach called Mochras), a meta-phorical stone's throw from the famous Cambrian strata, revealed an enormous thickness of Jurassic – which is otherwise not known on land for many miles in any direction. The Jurassic seas spread widely.

The Oxford Clay takes over a wide band of country running through Oxford to

Bedford and thence northwards to Peterborough and Lincoln. It marks the return to clay again as in the Lias, reflecting a return to a similar open sea. Collegiate Oxford is, of course, a grand essay in Jurassic stone; the humbler buildings are constructed of local limestones, but freestone from near Bath and elsewhere makes up the more finished buildings. The clay vale in which most of Oxford sits is followed by the upper reaches of the River Thames, ambling down from Cricklade, and lined with willows whose forgotten pollards now sprout extravagantly. This countryside is flat, and becomes flatter still as it approaches fenland south of The Wash. Older towns like Huntingdon and Spalding are attractive in local red brick, and there is even an eighteenth-century core to Bedford. Modern brickmaking techniques are peculiarly appropriate for using the Oxford Clay, which now accounts for something between one-third and one-half of all the bricks used in this country. The vast excavations one sees from the train near Peterborough are a twentieth-century counterpart to the fantastic holes excavated during the nineteenth century in the Cambrian mountainside at Bethesda in North Wales. Much of the new construction of the South-east grows out of these chasms: uniform, predictable, tested for strength, reliable, the Fletton-style brick is the Jurassic equivalent of Welsh slates. These pinkish-red bricks were not intended for facing, although they finished up on the outside of thousands of 'jerry built' houses, or else were plastered over with pebble-dash. Slates and bricks together constitute a kind of conspiracy against regional character, which shows in the uniformity of a hundred suburbias. There are now attempts to produce 'designer bricks' from Jurassic clays, which are an improvement aesthetically, but they, too, can ignore variations in geology which have made our islands so richly heterogeneous. The customer simply purchases the colour he fancies, and puts it down wherever he will.

One of the first brick pits I visited in search of fossils was in the Oxford Clay of Buckinghamshire. Washed out from the blue-grey clays were little ammonites in great numbers, each a perfect spiral an inch or more across, belonging to a dozen or more different kinds. Most were ribbed, and the ribs often bifurcated or trifurcated, but some carried great spikes, while others were as smooth as a discus. I looked them up in the great work by Arkell, and discovered that they carried names like *Cardioceras*. Particular species were recovered from one bed or another of the Oxford Clay, and thereby zoned geological time: in these pits you can see the flat beds laid one upon the other just as Arthur Mee had portrayed it in my *Children's Encyclopedia*. The ammonites were preserved in iron pyrites:

disulphide of iron, or fool's gold. This made them satisfyingly heavy, as if they had been cast in metal like a struck medal or a coin. Many had a golden glister. They looked and felt valuable. In the same quarry there were bits of fossil wood, quite black, but lacking the lustre of jet, which can be recovered from certain levels in the Lias. Jet was a coalified wood beloved of Victorian jewellers. In the Victoria and Albert Museum, there are some fine examples of elaborate jet carving in necklaces and earrings. It is said that jet found favour with Queen Victoria after the Prince Consort's death, being at the same time seemly and showy. Once, I found a fragment of a vertebra. From time to time whole plesiosaur skeletons are unearthed, and the brick companies usually allow them to be excavated – but fast!

One of the common finds in the clays of the Jurassic, like the Oxford Clay, is the fossil 'oyster' *Gryphaea*. This is one of those fossils that fits comfortably into the palm of your hand: quite precisely, a hand specimen. It is usually found preserved in a calcite which has taken on the black colour of the clay which enclosed it. The shell has a slightly flaky appearance, like a croissant. The shell of the larger valve of most species is very thick, often over a centimetre, and the fossil is accordingly rather heavy. The same properties make the shells very durable, and they last much longer than most fossils. They travel about. They get left over as a residue after erosion has done its worst, almost like flints. They become transported in rivers, or within the Pleistocene ice sheets. This is why *Gryphaea* can turn up in the oddest places: in the gravel on beaches, as a weathered pebble, even though the nearest Oxford Clay may be many miles away; in the garden, among the shards of crockery. It has been 'reworked' into younger rock formations than the Oxford Clay and hence might give a misleading signal of age to an unwary observer. I have seen a *Gryphaea* come tumbling out with a load of gravel bought from a Thames Valley pit (Eocene). Often, the species is *G. incurva*, which as its name implies is strongly curved. This must be the original for the vernacular name 'devil's toenail' . . . so here is old Beelzebub once again in geological guise. No doubt he sat upon the Devil's Chimney clipping his toenails. There are other species of *Gryphaea* which are flatter. All the species are supposed to have lived with the heavy, convex valve downwards, resting like a boat in the dark muds of the Jurassic sea floor.

Where the Oxford Clay reaches the Dorset coast it takes up the low ground behind Chesil Beach, and it is not surprising that such a soft clay does not make any natural exposure. Portland Bill projects imperiously into the sea, a rampart

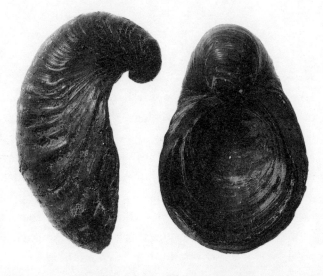

The Devil's Toenail, the clam Gryphaea, (actual size) one of the commonest fossils in the Jurassic clays. Why should the Devil be associated with such geological relics?

of tough limestone, and behind it the softer, older formations like the Oxford Clay slouch recessively. If the sea level rose, Portland would easily become an island. Chesil Beach is one of the largest storm beaches in Europe, more than seventeen miles long, and it owes its existence to the Portland prominence. The beach is made – almost entirely – of tough, rounded, flint cobbles, which endure while all else erodes. The pebbles decrease in size progressively away from the Portland end. The beach shelters the improbably long, thin tidal lagoon known as the Fleet, which is a site of international renown for wintering birds. Swans gather there in hundreds. The marine 'grass', *Zostera*, thrives in the Fleet, and forms the basis of a prolific food chain. Terns nest on the raw shingle.

The Jurassic rocks are folded and tilted in this part of the Dorset coast, from Chesil to Lulworth Cove and thence eastwards to Swanage, which removes the simple scarp and vale pattern. But there are limestones once again on the Oxford Clay, which give rise to a modest ridge along the main inland outcrop, and provide some humble, but attractive local building stones around Kingston Bagpuize, west of Oxford, and at Calne, Wiltshire. These limestones record the next shallowing of the Jurassic sea. They have long been known as the Corallian, and indeed corals can be found in some abundance at certain levels, along with other fossils of animals that liked life in a clear, warm sea. In detail, they are a remarkably varied set of rocks, including sandstones, the deposits of lagoons, and ironstones which were formerly worked around the delightful village of Abbotsbury, Dorset. These rocks can be examined along the Dorset coast at Osmington Mills, at Ringstead Bay east of Weymouth.

There is no better way to look at the top of the Jurassic than to follow the Dorset Coast Path from Weymouth to Swanage. From time to time the walker can dip down to the sea to sample the geology, but there is no need to take a hammer along to appreciate the scenery. Every dip and rise is the result of some change in the geology. The last clay interlude, above the Corallian, is seen in Kimmeridge Bay, when the Jurassic sea deepened one more time across southern England. These rocks are remarkable for being rich enough in oil to support a 'nodding donkey', which the coastal path passes, pumping up oil from its hidden reservoir. It has been working away since 1959, a kind of token Texas. On the beach at Kimmeridge, the rocks are not only clays, but also include shales and limestones, the latter making ledges on the foreshore. If you break one of these limestone beds with a hammer and sniff the broken surface, you will likely catch a whiff of a penetrating bituminous smell. This smell has been called 'oil man's perfume', and is said to bring tears to the eyes of even the most hard-nosed oil tycoon.

The limy ledges make a marvellously heterogenous habitat for marine life – this is Britain's first mainland marine reserve. Many of the species in Kimmeridge Bay are Atlantic species which are found no further east. At low tide there are pipefish, and hairy crabs and blue-rayed limpets; and as the tide drives you back to the shore, there is the fauna of a sea a hundred million years older lurking in the cliffs.

The rocks of Portland and Purbeck take the Jurassic to its close. Once again, above the Kimmeridge Clay, resistant rocks with many limestone beds recur, and it was this obduracy in the face of the worst the sea can do which has produced the face of this part of Dorset.

The peninsula of Portland is a defiant groyne thrown out into the sea. It is a strange, bleak place, dotted with quarries that have yielded what is probably the most well-known building stone in England – Portland Stone. This is a fine-grained oolite, a freestone, which can be cut perfectly flat, and which can be extracted in huge blocks. It seems naturally made to cope with monumental design. The size of the blocks is limited only by the spacing of master joints vertical to the bedding planes. The stone of all stones is from the middle of the Portland Formation, known as the Whit bed. It weathers practically white (contrast the clotted cream weathering of other Jurassic stones) and stands up to pollution rather well. Sir Christopher Wren used it for St Paul's Cathedral, and for many city churches built to replace those lost in the Great Fire of London

Portland Stone is the natural medium for public buildings: Ashmolean Museum, Oxford.

(1666). In Georgian Dublin, it was used to face many of the public buildings. It may be combined with red brick in formal designs. It was also exported to lend Old World dignity to Colonial Virginia. Oxford colleges added it to earlier Jurassic oolites, as in the facings of the Ashmolean Museum. One might say that we almost *expect* buildings with a certain public role, and classical design, to be built from Portland Stone, although few would know that this expectation was fostered in a shallow and tropical late Jurassic sea by the agitation of limestone particles. Now, the stone continues to be used for facing buildings – but cut into thin slices, where it can be added as a superior cosmetic to a mundane foundation. Even where city pollution has blackened it Portland Stone survives beneath its dirty patina, and it can be restored to whiteness when treated with high pressure water jets. It has what the Americans term *class*.

There are other Portland stones. The Roach has numerous fossils in it, inducing a curious arrangement of cavities, which can be used in a facing stone to great ornamental effect. In Perryfield quarry it is possible to find the distinctive fossil known as the 'Portland screw', a white, limy helter-skelter a few centimetres long; it is a mould of the inside of the spired snail *Aptyxiella*, the real shell of which has dissolved away. One of the youngest

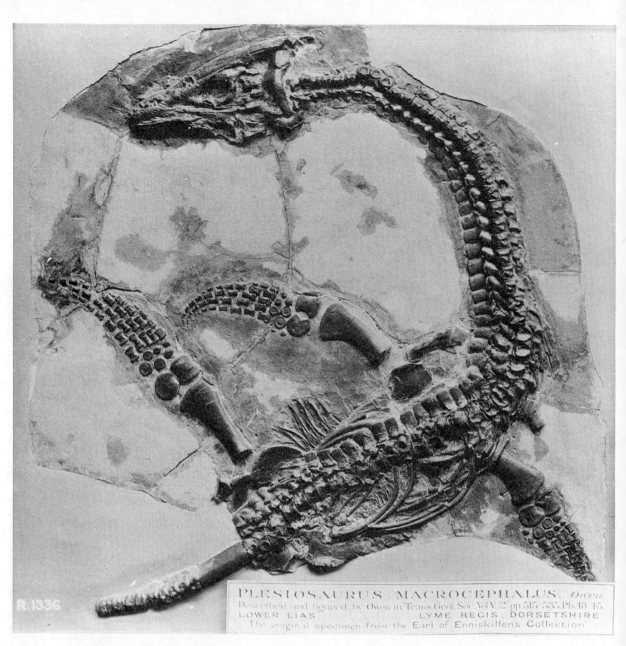

PLESIOSAURUS MACROCEPHALUS, *Owen*
Described and figured by Owen in Trans Geol. Soc. Vol.V, 2 pp.515-535, Pls.43-45
LOWER LIAS   LYME REGIS, DORSETSHIRE
The original specimen from the Earl of Enniskillen's Collection.

R.1336

The skeleton of a marine reptile from the Lias. Plesiosaurs range from the Jurassic to the Cretaceous in the British Isles.

Jurassic ammonites in Britain is a real monster, hence its name – *Titanites*. Particularly titanic specimens, several feet across, came from the Portland of Tisbury, Wiltshire, and are not uncommonly incorporated into walls or rockeries nearby. A fossil aficionado would covet them mightily. Being cast in Portland Stone they will endure indefinitely. They resemble the headgear of some monstrous mythical sheep, and there is no better simulacrum of the petrified horn of Jupiter himself (Ammon) after which all the fossils of this class were named.

The Isle of Purbeck defies the sea at St Alban's Head, where there is a lonely twelfth-century chapel dedicated to the eponymous saint. Purbeck is a would-be island, cut off by Poole harbour to the north, where the soft, young rocks of the Hampshire Basin are eroding away. Erosion will eventually bring about the final severance of the tough Jurassics from the mainland, when Purbeck will be an island in fact as well as name.

The Purbeck rocks are the youngest in the Jurassic of Britain. Perhaps the best place to see them is on the coast south-east of Swanage, or at Lulworth Cove. Limestones and shales record a further shallowing of the sea, and, eventually, a change to fresh water. In places you can see the fossils of mudcracks, baked by the sun of a Jurassic dawn, or the traces of salt crystals. There are the tracks of dinosaurs – the most everyday action of a great beast recorded in the indifferent mud. There are mammals; there are monkey puzzle trees, which are now confined in the wild to the southern hemisphere – our Victorian forbears developed a curious passion for planting them in the front gardens of houses with names like The Larches or The Limes; there are fossil beetles and dragonflies. The boles of trees are preserved in the Fossil Forest at Lulworth; tilted along with the beds that contain them they look like colossal petrified pots. The Purbeck rocks are a most important window into what life was like on land at the time – so much rarer than the record of marine life. I spent hours trying to find the remains of a woodlouse (*Archaeoniscus*) – a very rare early example of the only group of crustaceans to have made the transition on to dry land, and very successfully, too, if the number hiding under any dampish stack of wood is taken as a guide.

One of the Purbeck limestones is full of a tiny freshwater snail called *Viviparus*. This limestone produces the famous 'Purbeck marble'. It is not a true metamorphic marble of course but when it is polished, it becomes a dark and lustrous stone full of the white twists of ancient snails. Its virtues were discovered early; the Romans mined it. It is at its most beautiful in Salisbury Cathedral, where delicate columns surround the great stone piers, leading the eyes up to heaven to follow the fluted columns higher and still higher. The hands of twenty generations have felt its coldness, and thereby enhanced a polish which reveals snails that accumulated in shelly drifts one hundred and twenty million years before.

The steep folds along the coast to the west of the Isle of Purbeck come as something of a surprise in the South-east of England – where we expect the rocks to be tilted gently, at most. The folds on the Dorset coast are the westward continuation of the steep monocline at the heart of the Isle of Wight, which I shall describe later, in Alum Bay. The Lulworth Crumple is just that: a steep crinkling of the neatly stacked strata, best seen on the east side of Stair Hole. The resistant rocks of the late Jurassic lie seawards, a shield against the inroads of the waves. At Lulworth Cove the sea has breached these defences, and scooped out the famous bay in the softer early Cretaceous rocks behind them. It is a natural theatre in which the drama of the mortality of rock itself is enacted.

At Lulworth, the Purbeck Beds are tipped so that they can be readily examined; the inner side of the cove is a great cliff of Cretaceous Chalk, a rock which once again offers adequate resistance to the sea. This is one place where you can walk between two Mesozoic periods in as many minutes: from Jurassic to Cretaceous. The Jurassic rampart runs on westwards – to Durdle Door – but now you can see how the sea has won, as it always does. Looking westwards from Dungy Head a few rocks break the sea, the last gesture of the Purbeck Beds against their destruction. How long can the Chalk itself survive? There is no better place to reflect upon the vast stretches of geological time, how in this same place there has been land before, and that it eroded as the sea advanced and in the process made new rock, while now, again, even that later rock succumbs, to turn the cycle once more.

Lulworth Cove is burned on my memory. One hot, hot summer I spent a day there – shirtless – engrossed in the rocks. I was reflecting upon these cycles and more cycles of erosion and rock-making. But Life is different; its narrative does not repeat itself. Whether it 'advances' is debatable, but change it certainly does, just as the woodlouse in the Purbeck is not the woodlouse in my window box, nor the monkey puzzle tree the same species as grows in South America today. Nothing is certain in life except change and death; while rocks turn and return in cycles through the inexorable mill of the Earth . . . and all the time the whole of Lulworth Cove, and particularly the white, white chalk, was like a great parabolic reflector focusing and concentrating the sun's power. By the afternoon my back was glowing; when I walked into the cooling sea, I felt a jolt as the salt water met my fried skin. For a moment neither Jurassic nor Cretaceous seemed real, nor any history; all reality was dictated by the toasted tips of nerves.

13

The Weald

TUNBRIDGE WELLS lies near the heart of the Weald. Commuters pour from it into London; it is a safe and comfortable place, encapsulating the solid virtues of thrift and family.

Yet in truth this domestication is a recent phenomenon, based on the celebrity of chalybeate water as a health fad in the eighteenth century, and preserved by the subsequent expansion of the railways. Before the railway network pushed outwards from London, the Weald included many inaccessible areas; there are still jungles of little roads wandering and twisting through the countryside, and it is not so long ago that these were merely tracks, rutted deeply with the wear of centuries of cartwheels. Travellers in the early years of this century reported peasants with accents as incomprehensible as any in Cornwall. This has always been one of the most densely wooded parts of the country, and so it remains. There is still an aura of mystery in the rolling countryside. Much of the past survives: old buildings, ancient landscapes. On a misty morning the centuries roll away, and it is possible to believe that an Iron Age man rising early in the day might have greeted a scene little different in its essentials.

Much of what is seen is related to the hidden landscape. The geology of the Weald is quite straightforward, and it directly influences the landscape. North-south, passing through Tunbridge Wells, it comprises a generous anticline. The oldest rocks accordingly lie in the middle; framed by the Chalk – North Downs to the North, Beachy Head to the South – the earlier Cretaceous rocks underlying the Weald are arranged in their natural formations concentrically about the axis of the anticline. The scenery, broadly, follows the same concentric pattern. The coast also runs obliquely across the middle of the anticline, with Hastings near its centre. From Eastbourne to Folkestone the map tells us that we should see the succession of earlier Cretaceous rocks enclosed in the Wealden anticline, but in practice exposure at the coast is by no means continuous. The same succession of rocks are exposed on the south side of the Isle of Wight – and in its westward continuation, as I previously described in Purbeck and Lulworth Cove.

As one would expect in an anticline, the oldest rocks take up the ground in the centre. This is the land between Tunbridge Wells and the coast at Hastings. It is rolling, well-wooded country, probably the most attractive in the Weald. The Hastings Beds include several massive sandstones, which naturally tend to take the high ground: the High Weald. Clay beds between sandstones are sought out by the rivers and streams that intricately dissect the hilly country. The valleys they erode are often miniature gorges, flanked by bluffs of sandstone, as near Eridge and West Hoathly. Main roads tend to follow the ridges, while minor roads can duck and dive charmingly between banks and hedges in the valleys. There are deep woods, where sweet chestnut has long been naturalised, and where oaks are again achieving great proportions. The woods often flank the valleys, and can be dense, damp and matted with underscrub. At the coast, the sandstones make good cliffs at St Leonard's, near Hastings, and even more so eastwards at Fairlight Cove and Ecclesbourne Glen. Nearly horizontal sand beds can be traced for several miles here but, sadly, the best cliffs have the least sandy beaches, and where streams come to the sea they cut deep gorges and cascade on to the shore. This is truly wild scenery. The continuity of the coastal rock is interrupted for several reasons. There are faults that disturb and weaken the rocks naturally, and buildings which obscure them artificially, as around Bexhill and some parts of Hastings. Then there is the withdrawal of the sea within historical times around Rye Bay, which has left the cliff profile stranded, as it were, quite far inland. The Isle of Oxney was once more than a notional island, and the Rother Levels and Walland Marsh have not always been dry land, but were reclaimed over a long period going back to the Roman occupation. From a vantage point at Iden it is still not difficult to envisage the levels to the north as lying beneath a shallow sea, and who knows when it may yet return?

Winchelsea and Rye are two of the most interesting towns near the coast, both set upon sandstone heights on what is regarded as the old cliff line. Their geological eminence accounts for their being where they are. Some of the oldest buildings, like the ancient gatehouses of Winchelsea, or its gaol (now a museum), are built of the local sandstone, and very agreeable it looks, creamy and with an interesting texture; but it does weather badly. Younger buildings are constructed of local bricks, which are an attractive, warm red. But there are further traditions: hung tiles covering the upper half of the house are found all over the Weald, and there are numerous pretty houses following this pattern in these ancient towns. Then there is weatherboarding. This is particularly characteristic of Kentish towns like Tenterden, and inland East Sussex, in areas where, for no doubt good geological reasons, wood was more cheaply available than brick or stone. All these building styles can be found in different parts of Rye and Winchelsea, which are unfailingly interesting and various. Rye is a good example of how heterogeneity can enhance, and still remain true.

The older houses in Winchelsea retain huge cellars hewn out of the sandstone on which the town lies; these are a legacy of its days as one of the Cinque ports, when cellars would have housed barrels of sack, which had been traded for Wealden oak, or wool. The tops of the cellars just peep out at street level. Children like to imagine that they are dungeons, and a whistle echoes around them in a most satisfactorily creepy way.

By one of those delightful quirks that are typical of England, Rye and Winchelsea are the sixth and seventh of the 'Cinque' ports. The ports were established by the Normans to supply the Navy, and in return received trading privileges: the original five were Hastings, Romney, Hythe, Dover and Sandwich.

The wells of Tunbridge Wells are the reason for the town being there at all. The spring was discovered in 1606, and in the eighteenth century its virtues were highly regarded. The Pantiles are the legacy of the commercial development of the springs. Shops and inns are set about a generous courtyard; the verandahs of the shops together make a delightful covered walkway, reminiscent of the promenades in Italian cities. The Pantiles area is now unashamedly touristic but I believe it *is* still possible to taste the water. The springs are, of course, a wholly geological phenomenon, and Tunbridge Wells owes its very existence to its hidden landscape. The water at these wells is chalybeate, charged with iron salts. Iron is one of the essential elements for health, not least because an iron atom lies embedded within the complex molecule, haemoglobin, that makes blood red. Anaemia is lack of iron. It does not seem so absurd to recommend patients to drink from a chalybeate spring to cure what herbalists termed 'impurities in the bludde'. Modern science – always the spoilsport – has questioned whether iron can be absorbed into the system from the kinds of salts found diluted in minute quantities in springs. But it can scarcely have done any harm.

This is an appropriate point to consider briefly the subject of spas in general, which have only been mentioned perfunctorily before. Springs have been important in our cultural life since Roman times, and only lost their popularity in the twentieth century. Gazetteers, for the benefit of travellers in the seventeenth and eighteenth centuries are stuffed with references to the virtues of this spring or that. Bath, of course, is the definitive place for taking the waters. In its heyday, Society would migrate to Bath in the season, a fact which accounts for the elegance of the apartments and crescents there. It has even been claimed that this is where the *real* business of the nation was conducted at the time. The 'New Bath Guide' by Christopher Anstey (1766) gives an amusing account of the Blunderhead family set loose in this social minefield. Cheltenham and Harrogate were rivals to Bath, and both have claims to a matching elegance. Many a small fortune was founded on the discovery of a spring: a few cures, a medical endorsement or two, and then, most necessary of all, the patronage of somebody fashionable, and preferably aristocratic. It was possible for the more enterprising sufferer to make an intinerary which would take in several springs with different and complementary 'virtues'.

Springs must have a geological explanation, and the growth of Bath and its like is another expression of our underlying reality. The notion of health, or even wisdom, welling up from some unsullied spring has its counterpart in classical myth, and probably roots down into the Precambrian of our unconscious minds. At the very least, a bath in cold, clear water is invigorating, and in that way almost certain to be restorative. What is found in the water depends entirely on where the water has come from, and how it has travelled to get to the spring. Chemically pure water has very little taste, is even unpleasantly flat; those interested in proving this should try some distilled water used for car battery top-ups. It is dissolved salts of one kind or another which impart the taste. In the state of Nevada, the maps are dotted with 'arsenical springs', and we can count ourselves fortunate that most of the additions to the British springs are harmless, even if not potentially beneficial.

The commonest dissolved material is simply calcium carbonate derived from chalk or limestone – making 'hard' water. The source at Ashbourne, Derbyshire, or springs around Cheddar Gorge, or those near the foot of the Chalk, like Silent Pool, yield hard water of this kind, and it does taste good. Lime-rich water may have had a long, slow journey through limestone depths to reach the source. It emerges cool and free of contaminations. Magnesium often travels along with

Rye, Sussex. A charming mixture of all styles of tiles, bricks and weatherboarding.

Well wooded and undulating, the Central Weald viewed from Ditchling Beacon, East Sussex.

the calcium. 'Soft' waters will originate from lime-free rocks, such as granites, or sandstones. Dissolved iron salts in soft water are derived, as in Tunbridge Wells, from ferruginous beds, or else from peaty water, which concentrates iron. Waters enriched in iron often taste 'medicinal' – which may add to, rather than subtract from their reputation. Where the water derives from a deep source in the crust, it may be enriched in the last exhalations of the volcanic process; sulphurous, or fluoride-rich, or with a dash of manganese. In minute quantities, such additions impart a piquancy to the taste.

Water, it will now be appreciated, is a cocktail. Different recipes are more or less delicious. Human taste seems astonishingly sensitive to concentrations which are measured in milligrams, if not micrograms. There seems to be some accord about what tastes good; I have never heard one person pronounce the water delicious while another declares it undrinkable.

Oddly enough, the decline of the spa in this century may yet be reversed by the growth of the mineral water business. Entrepreneurs are already purchasing ancient springs. Is it low in sodium? Then it must be good for the heart! High in fluorine? Natural tooth protection! Nothing much in it at all? Its purity is exceptional! And so on. But it is good to see geological water coming back into its own.

There are several natural ways in which springs are born. Most often, they bubble up at the junction of permeable and impermeable formations – clays and many igneous rocks being the commonest of the latter. Where a permeable sedimentary formation overlies an impermeable one there will be a 'spring line' where the join is exposed at the surface – the northern edge of the Weald, where chalk overlies Gault Clay, is a typical example. A fault may produce a comparable juxtaposition. Or exotic waters may be released at deeper faults that pass down into profounder layers of the hidden landscape. Whatever their origin, we have reason to be grateful for the fashion for spas and wells; if it had never happened some of our most elegant towns may never have been built.

The stiff, blue Weald Clay surrounds the earlier sandstones which lie at the core of the Wealden anticline, and it forms a natural vale extending westwards from Romney Marsh. It is a region of fairly dense, but entirely rural habitation, and has the most impenetrable, criss-crossing network of country lanes. The M20 shoots by it, leaving the area largely untouched. Tiny, meandering streams dissect the landscape, in a pattern as complicated as the lanes.

All these Wealden rocks were deposited under conditions which geological textbooks always describe vaguely as 'non marine'. The early Cretaceous in southern England used to be described as having been laid down in a 'Wealden lake', and there is plenty of evidence for fresh water in places – beds entirely composed of the little snail *Paludina*, for example. Some of the sand beds were clearly exposed to the air as sandflats, and they carry the imprints of the dinosaurs that walked over them, distinctive three-toed tracks, left as casually by their perpetrators as any dog's on a beach. Examples can be seen at Fairlight Cove on the South coast. The modern interpretation favours the Weald basin as a kind of extended delta with the sea to the north, where marine deposits of the same age are well known. Sediments were derived from the west, the Cornish Peninsula and Wales, and from Armorica (Brittany and northern France) to the south, as well as from an area over East Anglia termed the London Uplands. It makes for quite a complicated geography in detail, but it was one which the dinosaurs must have known well, because their remains are more common in the Weald than anywhere else in Britain. They can be imagined picking their way between channels, now sprinting, now ambling across mud and sandflats.

The vegetarian *Iguanodon* has the distinction of being the first dinosaur to be described (1825). It remains the commonest dinosaur species in the Wealden rocks, although the graceful little *Hypsilophodon* and some carnivorous dinosaur fragments are also known. But the big discovery was made as recently as 1983, when the amateur collector, William Walker, found a large claw near the western edge of the Wealden outcrop, in the brickpit at Ockley, Surrey. The claw gave the dinosaur its popular name, 'Claws'. It turned out that almost the entire skeleton was preserved in the Ockley pit, which was an amazing discovery when it is remembered that a few, scattered fragments of any individual dinosaur are all that is usually found in the Weald. But the process of extraction of the skeleton and cleaning its bones proved extremely laborious, because the bones were preserved inside hard nodules. This matrix has been removed flake by flake in the nine years since the first discovery. The hero of this disinterment is Ron Croucher, the chief preparator at the Natural History Museum, who has devoted nearly a decade of his life to meticulous extraction of this wonderful animal. The first restoration of the dinosaur is now on display to the public at the Natural History Museum. Few of those who amble past the exhibit will have much conception of the sheer skill and persistence that has gone into showing this specimen to the world. The new specimen turned out to be even more exciting than was first supposed, for 'Claws' was a type of dinosaur which was hitherto unknown. It was formally

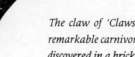

The claw of 'Claws' (Baryonyx) the
remarkable carnivorous dinosaur
discovered in a brick pit, at Ockley, Surrey, in 1983.

christened *Baryonyx walkeri*, thus guaranteeing its discoverer a deserved immortality. The dinosaur had comparatively large arms, equipped with its formidable weapon, which it may have used to hook out fish from the Wealden lakes.

One of the common fossils in the Wealden is the horsetail, *Equisetites*. Species of similar appearance still occur today in swampy sites, where they form miniature thickets of delicate and feathery branches, which are jointed, so that when you pull on a plant a little fragment bounded by a joint is all that comes away in your hand. They are primitive plants, without true leaves, and reproducing by spores. They have a history going back to the Devonian. Why, we might ask, should this humble plant endure, while *Baryonyx* and all its kind have passed from the Earth forever? Is it perhaps merely luck? Or maybe it is lack of ambition, to pursue a humble trade successfully rather than attempt to conquer the world. Anyone who has actually tried to *dig up* horsetails will know that they are, feathery look or no, extremely tough plants with an almost indestructible rhizome, and they can cope with dry banks as well as swamps. So I prefer not to accept explanations which hinge on Little Miss Humble passing unnoticed. Horsetails are survivors.

There are fossils of cycad-like plants as well, and monkey puzzle trees, and hints of plants with true flowers. The flora around the feathery horsetail swamps must have been quite rich, and certainly prolific enough to support browsing *Iguanodon*. Many insects scuttled about in the litter. From time to time there would have been a sipping sound as one of the fish took an insect which fell on the water. There were birds by this time, although their fossils are exceedingly rare; they may not have started singing yet. If so, this Wealden world would have been strangely quiet: no sound but the tramp of feet or grinding of teeth, perhaps the occasional reptilian hiss.

There are iron-rich nodules at certain levels in the freshwater Wealden. These were the basis for an iron-smelting industry which goes back to Roman times, and for centuries the Weald was probably the most important centre for this activity in all Britain. Because cutting trees, and other agricultural practices depend on iron tools, the geology of the Weald lay at the root of the domestication of at least part of the English landscape. It is easy to spot iron-rich intervals in a road-cut today because they weather to a bright rust colour.

The smelting of iron required three ingredients: the ore itself; a plentiful supply of charcoal for the smelting process, and water for power and quenching. The Weald was densely tree-covered, and remains one of the best wooded parts of the country to this day. Little streams were readily dammed to make 'hammer ponds', and a surprising number of these survive. You may come across one deep in a wood, with an obligatory resident moorhen pottering about, and surrounded by wood sedge and reed mace. The remains of a furnace by the pond can sometimes be made out, or there may be patches of the telling ferruginous stain. There are 'furnace coppices' marked on Ordnance Survey maps. The steep-sided, miniature gorges so typical of the Weald were ideally suited to damming up a series of ponds stepping downhill. The charcoal was derived not from great oaks but from coppices or pollards that were specially trimmed and husbanded for the purpose.

Weald iron is low in sulphur and very durable, and there were more than a hundred small furnaces by the sixteenth century. The last furnace closed in 1828. There is some debate about the effects of these furnaces on deforestation. Did voracious forges gobble up trees in an endless hunger for charcoal or were the furnaces sustained by careful management of the nearby woodland? The large oaks were used for the spars of ships rather than for fuelling forges and the steady decline in iron smelting was more likely caused by the development of Carboniferous ores near the coalfields, rather than deforestation. Much richly wooded country in the Weald survives, even though it must be less forested now than in the days when the Saxons ran their pigs into an impenetrable wilderness to fatten up on wild acorns. Perhaps, for once, we have reason to be grateful to an industry for the survival of the countryside.

The Ashdown Forest is different from wood-and-valley Weald. It is heather-clad heathland with birches and pines, interspersed with valley bogs where sundews grow among sphagnum moss. There are adders here, and stonechats hop in and out of the gorse bushes. Such open country is found on the more porous sandstones, where soils are thin, and nutrients are easily leached away by heavy rain. It is reminiscent of the true heath that crowns the Greensand summits to the west, and it is first cousin to the New Forest itself, grandest of southern forests.

The sea returned to flood the Wealden basin and to inter dinosaur bones and horsetails alike beneath new floods of sands, but they were sea sands this time. Above the Weald Clay, the Lower Greensand produces a ridge of hills that runs, with a few breaks, from Hythe and Lympne on the Kent Coast to Ide Hill in west Kent, and thence westwards again to Leith Hill and the further Surrey Hills. The sandy rocks are called the Lower Greensand, which is a confusing name, because they are seldom green. Instead, they are usually ochre or rusty, the colour of an ageing Italian *palazzo*, or Van Gogh's cornfields. Heights of 200 m or more give the Greensand hills sufficient eminence to afford some of the best vantage points from which to survey the Weald, and to appreciate how its geology has sketched its features, how clays make the low ground a chequerboard of fields or damp woods, how sandstones take the ridges, bare or wooded by turn. North from Leith Hill there is the younger scarp of the Chalk Downs at Box Hill, where the alkaline soil makes almost every plant a different species from those on the sandstones. Pine and birch trees, and the heather family, and harebells and wiry grasses clothe the greensand. And there is gorse, just a yellow-flowered pea armed with prickles at the expense of leaves, for which the great Linnaeus is said to have fallen to his knees in gratitude to God. Somewhere in a gorse patch there will always be flowers, whatever the month. In spite of the encroachment of greater and still greater London there is wildness still in the Surrey Hills. The National Trust has preserved several of the Greensand heights. In the summer, it is still possible to lie comfortably on springy turf and hear the rasping of crickets and the soporific buzz of bees about their business.

I always associate Lower Greensand country with holloways. These are sunken roads or tracks enclosed deeply between sandstone banks. The trees on either side arch together to meet above and blot out the sky; even on a sunny day all you see on the ground is a random dappling of gold coins which the sun drops through the canopy. Greensand banks sometimes form miniature cliffs, and dark crevices support groves of male fern and patches of creeping liverworts; ivy hangs down in curtains, a place for wrens and dunnocks. Holloways are usually ancient routes worn down by the passage of feet and hooves, cart and carriage. Then too these lanes will bend this way and that out of respect for parish boundaries – or perhaps out of sheer perversity – making any calculations based on the flight of crows absurdly approximate. Holloways are secretive and enticing places.

It is likely that the trees on the banks did not always leap so high; they may be hedges long since 'grown out' following the abandonment of regular laying. Former hedgerows often reveal themselves by the trees having preposterous dog-legs before their branches grow upwards; the horizontal part is the legacy of the last time the hedge was properly laid. Dr Hooper has alerted us to the fact that ancient hedges are rich in species of hedgerow trees and shrubs, and a test of the antiquity of the holloway is to count the number of different species found over a thirty yard length of the flanking 'hedge'. Although the rule of 'one species per hundred years' age' has been criticised for offering a spurious precision, a record of half a dozen or more species over a short length is still a good and reliable guide to antiquity. When we learn that many hedgerows are six hundred, even a thousand years old the cavernous depth of the holloways is not so difficult to understand. Only tarmacadam now arrests the abrasion wrought by time.

Between the Lower Greensand, which may not be green, and as we have seen is often rusty with iron, and the Chalk, which is definitively white, there is the dark Gault Clay and the Upper Greensand, which, surprisingly, actually *is* green. The Gault Clay forms a vale not merely in the Weald, but also extends beneath the Chalk all the way from Lyme Bay to the Wash, and eastwards into the Isle of Wight. The open sea spread widely by this stage in the Cretaceous, overstepping westwards on to progressively older formations, until in south Devon the Gault rests on the eroded Triassic, Permian and, eventually, Carboniferous. This is one of the great drownings of our islands, and its legacy on the geological map is one of the major unconformities. The Upper Greensand occasionally makes a feature, most notably near Selborne (Hants), where it reaches 180 metres; it often grades upwards rather gradually into the Chalk. Along the Vale of Pewsey it lines a richly agricultural valley slumped between the heights of the Marlborough Downs and Salisbury Plain. The soil produced by the weathering of the Upper Greensand is a true loam, and much more fertile than the sandy soils produced by the Lower Greensand. You have only to look at the dense web of old villages along the Vale of Pewsey compared with the emptiness to north and south to see that farmers appreciated the geological reality in bushels and pecks, without the need to consult a map of the formations.

The Gault is a black-blue clay and very sticky: it makes little natural exposure, but it has been much excavated for bricks. Beware its adhesive power. I have lost single Wellington boots in pits at Devizes and near Dorking through underestimating its grip. The Gault is unreliable stuff; it engenders land slips when it is loaded, as it does west of Lyme Regis, at Folkestone, and in several sites on the Isle of Wight. The Gault signals a return to those stiff, 'Jurassic' clays such as the Oxford Clay, and is likewise full of ammonites and belemnites and clams. The most common ammonites used to be called by the name *Hoplites*, referring to those Greek warriors who fought in phalanxes, and indeed some of these ammonites are well armed with spikes. The textbooks also show strange ammonites that have become 'unwound' into weird hooked shapes, or open spirals. They were alleged to show almost pathological deformity of design, presaging the demise of the ammonites as a whole, which happened somewhat later at the end of the Cretaceous. This curiously fatalistic view of evolution – powered from within – has been replaced in recent years by a kind of counter version in which life, poor evolving life, is seen as a victim of world, or even extra-terrestrial, events; luck alone drives the wheel of fortune, and survival may be a matter of chance. A real relativist would probably say that the nature of explanation itself depends on the state of our human society. Only the ammonites entombed in their treacherous, dark mud are released from the obligation to mean whatever the observer wishes them to mean.

The Weald provides beautiful building materials. Stone is often sandstone from the non-marine centre of the anticline: Weald Stone, or some of the tougher sandstones near Hastings, or the brownish sandstones used around Tunbridge Wells. Old churches in East Sussex are dignified by pleasant, creamy-fawn sandstones, which I have not matched exactly in any other part of the country. Bodiam Castle is constructed of huge blocks of sandstone which makes it seem truly impregnable. It looks an improbably belligerent place, moated, castellated, and massive, as if it had been built to the instruction of a paranoid laird, haunted by memories of Border skirmishes. When the River Rother was open to the sea the fortification would have looked less incongruous. The other stone is the Kentish Rag, a local development of the Lower Greensand, a tough impure limestone, which was a favoured stone in medieval London. The White Tower at the Tower of London is built of it, with additions of limestones brought in from Caen, an import which was not at all unusual at that time, when political and commercial links to northern France were much closer. Knole is an ancient stately home near Sevenoaks, Kent, which is constructed from a patchwork of Kentish Rag, which gives it a kind of roughened grandeur.

But perhaps the Weald belongs most to bricks and tiles. There was an abundance of local bricks from its several clays: good and durable bricks of warm reds, but varied enough never to be oppressive. The addition of hung tiles in the upper half of the houses just adds to the delight of it all. In villages like Burwash (East Sussex) and towns like Rye or Lewes no two hangings are the same. Hanging tiles are protected by the projecting eaves, so that they stay as bright as when they were made. Herstmonceux Castle is entirely made of bricks, and although it is moated and castellated like Bodiam it feels altogether tamer. In the Weald, even some older houses with timber frames often had bricks, rather than wattle and daub, as the infill between oak beams. In Kent, bricks and tiles were put to good use building oast houses, which were used for drying hops, a crop which thrives in the Kentish soil. As a child after the war I remember travelling through mile after mile of strings, which the hops climbed like runner beans. There are few hop fields now. But the oast houses remain, often tucked into intimate hollows so that from a distance their conical towers look like a strange place of worship. My children used to call them 'ghost houses', and so, in a sense, they are.

The methods of manufacture of local bricks had another benefit: the ends of the brick were frequently of a different shade from the sides. The style of laying bricks with alternate 'stretchers' (sideways) and 'headers' (endways), means that chequerboard effects are introduced just because of these colour differences. In eighteenth- and early nineteenth-century houses this was used to great ornamental effect. The Weald is full of good examples, but most of the older brick-built town centres in England will have them, especially among larger Georgian town houses which give straight on to the High Street. The modern habit of laying nothing but course after course of stretchers (because it saves bricks) is another capitulation to uniformity in the interests of economy. An older economy was employed in the Weald, and especially Kent, in the use of weather-boarding rather than bricks; this can be most attractive if it is done well. Some weather-boarded Kentish villages are reminiscent of the older parts of New England, or should it be the other way around?

Not all bricks produced in the area are red or brindled. The Gault Clay, near

Maidstone, produces 'whites' (cream or yellow, actually) much employed by Victorian builders. Near Faversham and Sittingbourne the Thames Valley 'brick earths', originating from clays laid down in the valley in the Quaternary, have produced the 'London stocks' which have built large parts of the capital, especially in the Georgian period. These have a distinctive yellowish-brown colour, tending towards the latter under the influence of pollution. They last well, and even now are being buffed up sparklingly in the new, cleaner London.

The hidden landscape in the Weald controlled the shape of the countryside, the course of its development, its industry, its vegetable life, the buildings that adorn it – almost everything, it seems, except the weather, and the urban history. Yet does one walker in a hundred think of what lies beneath his or her feet, other than occasionally picking up a stone, and hurling it away for the dog to chase?

The rocks of the Weald come to the surface again in the south of the Isle of Wight, as a consequence of the sharp fold running through the centre of the island. You can see a lot here within a dozen miles. The Wealden non-marine rocks are exposed on the shore west of Atherfield Point, which is a good place for those interested in fossils. At Shanklin, a stream cuts down through the Lower Greensand to make a precipitous little valley tumbling down to the sea, one of the island's famous chines. Within a few minutes you are away from cottagey tourist shops selling knick-knacks and pot pourris and into a shady and verdant chasm following a path that dodges this way and that. It must have been well-nigh impassable before the path was made. The greensand genuinely appears green here, but from a patina of algae and moss. In one of the sunnier spots I remember seeing Wild Madder (*Rubia peregrina*), a kind of coarse goose-grass, for the first time. Perhaps it still grows there.

South of Shanklin Chine, the Undercliff path to Ventnor weaves charmingly up and down under overgrown beeches and sycamores, dodging in and out of sun and dappled shade as it goes. The sea lies far below. It is all rather overgrown now. Like Ventnor itself, it seems to have seen better days – the rough wall that guides you has tumbled down in places, and some of the steps are a little treacherous. Nonetheless there is a certain delight in this genteel decay: the whole route is so overgrown in places it is like being in a tropical jungle, with moss and fern and the distant calls of unseen birds. From time to time blocks of rock emerge from the growth, and this is where the puzzle starts. The walker with even a smidgin of geological knowledge will recognise well-bedded, rubbly sandstones, but the bedding is far from horizontal – indeed, it can point in any direction. The truth is: they are tumbled blocks. The real outcrop lies landward of the path, forming a cliff which can be glimpsed through the trees, and there the bedding is close to horizontal. These cliffs are in Upper Greensand. The most prominent of the fallen Greensand blocks is the 'Wishing Throne' by the path, where the bedding is truly vertical. The landslips cover all the ground between the path and the sea. The culprit is that unreliable, wellington boot gobbler, The Gault Clay, which lies beneath the Upper Greensand and caused it to founder.

The path wanders on until it reaches Bonchurch, where the poet Algernon Charles Swinburne was incarcerated with his minder in a house which presents to the world nothing but intimidating gates. Those who do not admire his verse would say that it was no more than he deserved. The tiny old church at Bonchurch is built of the local sandstone, and is almost indecently picturesque. It is hard to believe, now, that this coast had a reputation for smuggling.

I always associate chines with the hart's tongue fern (*Phyllitis*). On the Isle of Wight it grows in whole drifts where nothing else will. Its leathery leaves make it look more like an herbaceous perennial than a fern, but turn over a leaf and the reproductive organs on its back reveal it for what it is. Fossils like it date back several hundred million years, and so its ability to tolerate shade – even the disagreeable canopy provided by sycamore trees – must have stood it in good stead. I once found one growing in an old well in the middle of the Chalklands, miles away from the nearest fernery. Such is the advantage of spreading by means of spores. In damp chines, hart's tongue ferns are found along with liverworts, which are little more than creeping, photosynthesising pads; they must have been running over damp sandstones in this way since the Devonian. In the same places there are a few flowering plants, like the wood avens, *Geum urbanum*, which the herbals tell us is effective against all manner of intestinal and stomach complaints: 'inward fluxes, other burstens and ruptures, wind colics'.

From the ferns of the damp undercliff glades it is a short, but steep walk upwards to the higher Cretaceous: across the main road at the top of the cliff, and you are on a Chalk hillside: marjoram and thyme will greet you, and you will hunt ferns in vain. Time and rock have changed, and with it almost all the flora. To climb to the Chalklands is to go to a different world, for the changes to the humblest herbs mirror changes that happened a hundred million years before.

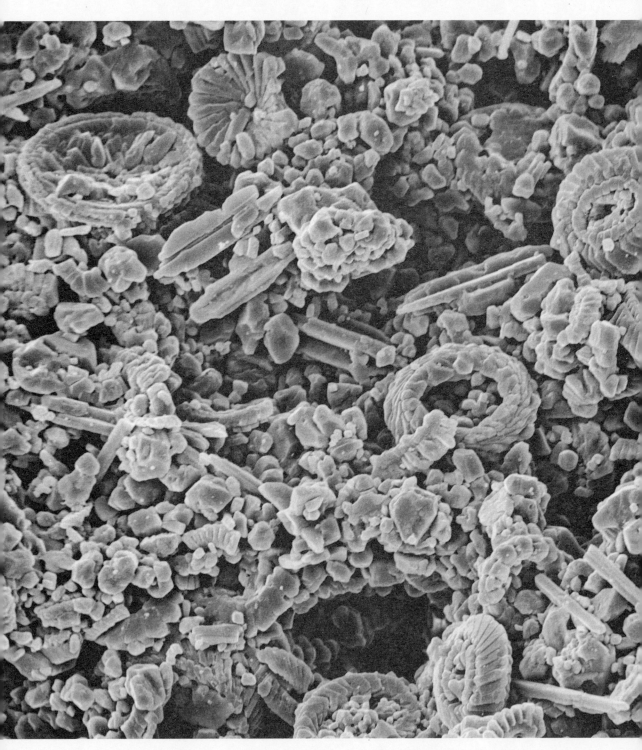

The ultimate components of the Chalk are these minute, and beautifully summetrical coccoliths, secreted by algae. To be seen clearly, these must be photographed by the electron microscope at several thousand magnification.

The Bog Asphodel, a plant of acid bogs, such as those developed on the Millstone Grit.

The 'New Red Sandstone', crossbedded sandstone of typically bright colour found in South-west England.

Left   *The limestone rock known as an oolite, so prominent in the Jurassic limestones, is composed of small, spherical grains. This is a magnified view of an oolitic limestone.*
Below   *Bath: the definitive stone city, employing Jurassic rock to monumental effect.*

Facing page   Top *Chesil Beach, a magisterial sweep of cobbles, viewed from Portland Bill, the natural groyne which protects it.*   Below *Stairhole Cove, Lulworth. Clays and limestones of the Purbeck Series are splendidly seen in folds which seem surprisingly intense in the gently folded South-west.*

Above   *The Red Chalk at Hunstanton, Norfolk, is another aspect (facies) of the familiar white rock.*   Below   *Plants on Chalk. Bee orchid (left) may be abundant one year, and almost rare the next. The uncommon and delicious morel (Morchella esculenta) (below left) is a fungus which appears in spring in the Chiltern Hills. Wild thyme (below) is common, especially on anthills.*

Facing page   Top   *Sunken lane in the Weald, etched by the passage of countless carts. Thorncombe to Godalming road, Surrey, in Hythe Beds, mostly sands with chert bands.*
Below   *Geology and scenery, near Box Hill, Surrey. The low ground beyond the main road is occupied by the soft Gault Clay, the ground rises into the more resistant Upper Greensand beyond the railway, and further again into the Chalk of Box Hill (North Downs) on the left.*

Above   *Alum Bay, Isle of Wight. The famous coloured bands of the early Tertiary rocks have been folded so that they lie vertically above the Chalk.*   Below left   *The wild Gladiolus,* Gladiolus illyricus, *which can be found in favoured sites in the New Forest.*   Below right   *Nummulites. These are among the largest of single-celled organisms, and abounded in the early Tertiary (Palaeogene), although only found at certain levels in England. This species is* Nummulites gizehensis, *a species which contributed to the limestones of which the Egyptian pyramids were constructed.*

Facing page   Top   *Greenstead Church, Ongar, Essex. Oak tree boles have produced the walls of the oldest parts of this Norman church — the geology of this part of Essex yielded no good local stone for permanent construction.*   Below   *In the absence of good building stone, cottages can be literally cobbled together from any likely materials that come to hand, as in this example from Norfolk.*

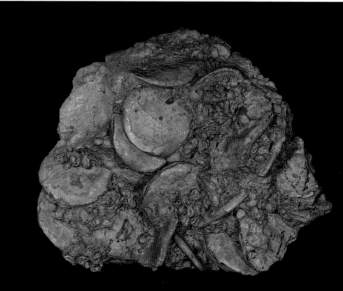

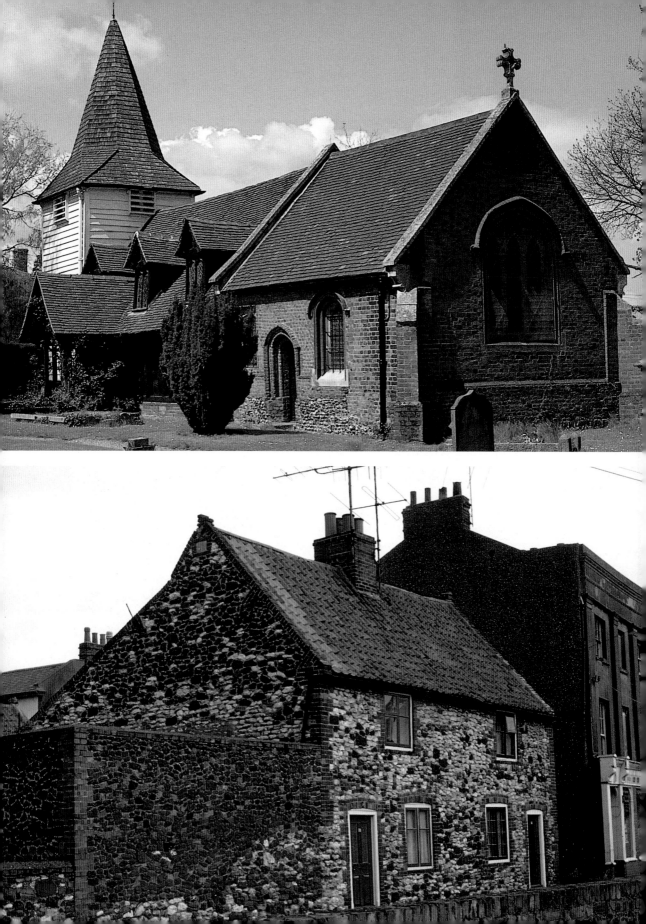

Fingal's Cave, Isle of Staffa, West of Mull: the other, dark volcanic face of the Tertiary. Lava flows solidified into basalt columns of remarkable regularity.

14

The Chalklands:
Downs and Flints

CHALK: COMMONPLACE stuff at first glance, but chalk is special. In no other part of the geological record are the connections between landscape, flora, farming, history and the rock beneath so clear as in the case of the Chalk. The strata comprising the Chalk are readily recognisable. On the geological map it is one of the most distinctive formations, easily mappable, and recognisable from a distance, even from a passing car. The Chalk is our typical late Cretaceous deposit, and represents a momentous time, for at its end the dinosaurs passed from the Earth forever, as did ammonites, and a host of lesser organisms. After Chalk times, we see the beginnings of the modern world.

The Chalk is a white, chemically pure limestone. In many places it is also rather soft for a limestone, and can be scratched with a finger nail. Compare it with the Carboniferous Limestone, for example, some parts of which are equally pure, and it seems positively crumbly. You can nibble chalk if you have acid indigestion and it will make you feel better fast; white lozenges sold at great expense for the same purpose are often nothing more than chalk with a little flavouring. The calcium carbonate of which the chalk is almost entirely composed reacts with the hydrochloric acid of the stomach to reduce acidity. You can see the same effect if you sprinkle a little vinegar on to a lump of chalk: it will hiss and fizz until the acetic acid is entirely neutralised. The same rock type is known from other parts of the geological column – and this is lower case 'chalk'. *The* Chalk with a capital 'C' is our late Cretaceous deposit, uniquely thick and widespread.

It was on the Chalk that I first learned the excitement of discovering fossils. They are not always common, but can be unusually beautiful. Some parts of the

Chalk, particularly the Upper Chalk, do not yield their secrets easily. I have spent hours in a quarry with little to show for it other than a crushed brachiopod. But at other times there can be a magic in your hammer that persuades the Chalk to part with a dozen sea urchins in half an hour. The fossils have a slightly pink cast, or you spot the rotund end of a sea urchin projecting from an otherwise smooth face of rock. It is usually not worthwhile sitting and smashing pieces of rock, as the scattered occurrence of the fossils makes for a poor chance of success. The Chalk hunter rather crawls crab-like and head down over fallen blocks, or peers closely at cliff faces looking for a hint, a tell-tale sign, of something he can take home to dig out. But the Chalk is full of chimeras which mislead. Flints are the worst, for they often mimic what you most wish to find, and they may be scattered on the chalk face just like the real thing. Occasionally a flint cast is a fossil; those of the heart urchin *Micraster* are extremely durable, and may turn up in unexpected places far from the Chalk outcrop, even in London gardens. Such heart urchins have survived the weathering down of the Chalk that once enclosed them, have probably been incorporated into a river gravel, and may have been reworked during one of the glacial periods, before eventually turning up on a garden fork in London W12. So resistant is flint that it endures unchanged except for changing brown from its incarceration in clayey subsoil.

A freshly collected heart urchin fits snugly in the palm of the hand like a small potato. The feel is slightly rough to the touch because it is covered in tiny tubercles that, in life, would have held minute hairy spines. On the top surface there are five sunken grooves. The number five shows at once that this fossil belongs to the echinoderms, which alone in the animal kingdom adopted five-rayed symmetry. There are other sea urchins in the Chalk; some of them have huge, club-like spines, and it is much more common to find the spines than a whole animal. Few of these fossils can be collected in perfect condition, but chalk is soft enough to make the discoveries relatively easy to clean. In many cases an old-fashioned bristle toothbrush can be used to brush off excess chalk (I do this under water) until the calcite surface of the fossil is reached. If you are very careful in this cleaning you will discover the kinds of fossils which grow upon other fossils. The sea urchin test is often covered with the minute colonies of creeping sea mats (Bryozoa) of which there are dozens of species. A hand lens will reveal this miniature world. There is a vast number of species of many larger fossil types known from the Chalk: clams, snails, ammonites (in certain places only), sponges, brachiopods, starfish, sea lilies, crabs, corals (rather rare), belemnites. Some particularly exciting fossils can be quite common in the right locality, like triangular shark's teeth still sharp enough to cut your finger.

But now imagine that all these larger relics of life have been removed from a lump of chalk: look closer. At higher magnifications ($\times 50$) the 'washings' from the Chalk show tiny, chambered, calcareous shells. These are the tests of foraminiferans, single-celled animals which still thrive in the oceans today. A millimetre or so in diameter, these tiny fossils teem in millions beyond the powers of computation in the Chalk, so much so as to make up much of the rock in places. The smaller ones are likely to have been part of the plankton – the floating population of the sea. These often have inflated chambers, and may be coiled like diminutive ammonites. There are other shapes as well, top-shaped tests, or tiny bunches of 'grapes'; some bigger ones may have been sea bottom dwellers along with the clams and urchins. And now let us imagine that we have picked out all the foraminiferans from the remnant of our chalk sample, together with tiny fragments of shells (those of the clam *Inoceramus* are particularly abundant). There is still a fine white powder remaining, as fine as refined sugar, and much the same colour and texture. To examine this we need to go to even higher microscopic power. If you turn up the light microscope to its highest power some of this dust appears to resolve itself into little wheels: circular, perhaps slightly elliptical, under polarised light they show curious cross patterns which surely implies they have structure. To really see what these little wheels are like it is necessary to use an electron microscope; these objects are only a few thousandths of a millimetre across. And only now is it clear that they are minute rosettes of calcite crystals, stacked and splayed like fans of playing cards. Despite their minuscule size they too are fossil remains with order and design. They are coccoliths – calcite plates secreted by minute, planktonic marine algae. Like planktonic foraminiferans, they also have modern analogues, which is why it is known that a single cell secretes many such plates. They armour the cells, ten or more at a time. We can only marvel at the incredible numbers of such algae there must have been to contribute to the Chalk. And since algae require sunlight to grow there is a sense in which the brightness of the Chalk preserves the sun that once played on the Cretaceous seas.

So now the astonishing truth is clear. The Chalk *is* fossils, virtually in its entirety. The cliffs, the downs, and the quarries are the construction of organisms in their trillions, even down to the very dust. Its whiteness, its purity, is the pure calcite manufactured by animals and plants, and here unsullied by sediments from other sources.

To most observers the Chalk is remarkably uniform in appearance. The White Cliffs of Dover seem blockily implacable, which partly accounts for their symbolic value in time of war. This appearance of uniformity is deceptive; once you 'get your eye in' all kinds of subtle differences become apparent. Nonetheless on the Kent coast around Broadstairs or Dover the cliffs show a few bedding planes running horizontally along which the sea spray erodes a little more readily: you can usually follow a single bedding plane as far as the eye can see. At right angles to the bedding planes are joints, and when there is a cliff fall the joints will serve to define the massive block that comes tumbling down to the shore.

The chalky cliffs at Broadstairs must have been the first rocks I ever saw, as this is where I spent my childhood holidays. I have no geological recollections. This was a time when ordinary people like us spent their holidays in guest houses ruled by landladies more implacable than a chalk face. I do remember sitting rigidly at the table gloomily contemplating a mountain of pallid cabbage. Margate then boasted lights that were supposed to rival Blackpool's; perhaps they still do. But in these distant impressions the Chalk serves only as a backdrop to the stage of the seashore.

Through the Chalk run lines of flints. Usually these outline and follow the line of bedding, the Cretaceous sea floor, rather closely. Sometimes flint beds form more or less continuous sheets, or they may be dotted here and there. Some of these flint beds can be traced continuously for miles. Flint is very hard, and is chemically unrelated to chalk, being a form of silica (silicon dioxide). Silica is therefore the other ingredient besides calcium carbonate in the Chalk Formation, and being so tough and inert it alone survives from the Chalk once the white rock has been weathered and eroded, with profound consequences for our landscape. The formation of flint within the Chalk was argued about for many years but is not now such a contentious issue. The source of the silica was certainly organic, like everything else in the Chalk: for some organisms use silica to build their skeletons rather than calcium carbonate (calcite). There are common siliceous sponges in the Chalk, and the sea then likely swarmed with single-celled radiolarians which had built delicate silica skeletons for hundreds of millions of years. The kind of silica of which they were composed is analogous to opal, and is not particularly stable. It dissolved soon after the dead tests of the animals had fallen to the sea floor, only to be redeposited as nodular layers within the chalk

Heart Urchin (Micraster), *a sea urchin that probably burrowed into the soft chalk sea floor. Fossils of* Micraster *have been called 'shepherd's crowns'. (About twice the natural size.)*

Freshwater Bay, Isle of Wight. Well-bedded Chalk, the bedding planes often outlined with flints. Tennyson often used to walk the cliffs here.

sediment – as flint – in a more stable form. Fossils often acted as the nucleus for this redeposition, as you will see when you find a sponge within the rotten nucleus of a flint. Because some sponges are spherical, those flint nodules which formed around them are spherical, too. Some of the regular enquiries at the Natural History Museum concern 'fossil balls' – spherical flints formed around sponges and then picked up long after the chalk that once enclosed them has been worn away. You can pick up balls like this on the ground of beechwoods in the Chilterns. Are they, perhaps, stone cannon balls? Most of these balls will shatter if bashed with a hammer, and inside will be found the crumbly remains of a Cretaceous sponge.

Because flint is deposited somewhat haphazardly, the flint nodules can assume various shapes. On the East Anglian coast circular ones that resemble knobbly life-belts were once named *Paramoudra* as if they were real fossils. They still attract attention as interesting objects on rockeries. Flints provide the commonest of false fossils, or pseudofossils, and some of them are spectacular. Passable 'faces' turn up from time to time. Could this have been the source of the legend of the Medusa, whose glance could turn men to stone? 'Feet' are commoner, those with five toes rather rarer. I have seen horses' heads, and a mermaid that would be unlikely to inflame even the most desperate seaman. The prize must go to the visitor to the Natural History Museum who brought a pointy object wrapped in a pillow case. 'Has one of *these* ever been found fossilised before?' he asked. Whisking off the pillow case, he revealed a rather larger than life size set of male genitalia.

The homogeneous chalk produces a landscape of swells and broad valleys: no crags like other limestones, nor sinkholes, nor waterfalls. There are not the weaknesses and differences within the Chalk that erosion can seek out to make the landscape stepped, or for the bare bones of rock to burst through. Chalkscapes are, I suppose, an acquired taste. So much has been ploughed up now, and put to barley, where once there were flocks of sheep. The proximity of the rock to the surface is shown by patches of creaminess after ploughing.

Indestructible flints speckle the surface of the soil. Fertiliser must be applied to make it productive, a process that removes many native weeds and herbs. It is awkward soil to walk over: large flints twist your ankles and sharp ones cut your children's legs. Cattle are now grazed on the steepest slopes, and the ruckles of their tracks score the hillsides; this may sometimes mimic stratification, so beware false geology. Yet the chalk landscape has about it a sweep and openness which is attractive. The contours are like those of the human body, the swell of buttocks or breasts. Some of the gentler of Henry Moore's reclining figures might be cast from this landscape. When the wheat and barley is ripening, the pattern of swathes of subtly different ochres almost makes one forgive the lack of hedgerows, and there are still occasional patches where poppies have escaped the weedkillers to splash the hillsides with scarlet. Poppy seeds have extraordinary longevity: they survive for many years in the soil, and their germination is triggered by light. Hence their apparently mysterious reappearance when the plough glances into a hedgebank undisturbed for years.

Where the Chalk rises above the Upper Greensand underlying it, it can make real height. The Marlborough Downs run alongside such a greensand valley, and further west lies the Vale of the White Horse; precipitous scarp slopes make these hills seem much higher than they really are. Parts of the North and South Downs rise as dramatically. The 'Iron Age hillforts' like Wittenham Clumps (Oxon) and Wayland's Smithy were frequently sited on top of the slopes, and very sensibly, too, for no matter how different the countryside was at that time, the view from the top would have been as panoramic as it is now, and the approaches as well protected. The chalky soil is thin, and there is evidence that it was cleared of scrub by our ancestors when other parts of the country were still thickly

The White horse at Westbury, Wiltshire. Graffiti using natural materials.

wooded. Parallel ditches and obscure depressions are what one sees of these hillforts as you trudge and scramble round on foot. The one near Inkpen Beacon on the Marlborough Downs has a gibbet nearby which stands on the hilltop, visible for miles.

The usual pattern in Chalk outcrop is for the scarp slope to be steep, the dip slope more gentle. But occasionally the Chalk forms a ridge which is steep on both sides, as on the Hog's Back near Guildford (Surrey). Chalk heights made natural sites for roadways long ago: thousands now walk The Ridgeway, following lines laid out by drovers and itinerant traders for hundreds of years. Thus the outcrop of the Chalk set out some of the most important historical routes across the country.

Where the chalk lies horizontal and has its most extensive outcrop is on Salisbury Plain. Here you can see further than almost anywhere in South-east England. It is almost steppe-like. It cannot be coincidence that this was the last habitat in Britain of the Great Bustard, a giant, shy bird requiring clear spaces, which it now finds only in the steppes of eastern Europe. Nobody forgets their first sight of Stonehenge across this clear space. The army values the same perspective for manoeuvres, but we should be grateful for this has saved areas of Salisbury Plain from falling under the plough. Most of Salisbury Plain is underlain by very white Upper Chalk. The flint does not lend itself to making extensive stone walls, so there are not even the periodic enclosures of the kind that break the outline of the Gloucestershire plains underlain by Jurassic limestones.

Chalkland, and particularly Salisbury Plain, is barrow land. The map is dotted with tumuli, which despite the passage of two thousand years still contrive to look unnatural. The burials they once contained have almost invariably been exhumed. The common round barrows and larger long barrows follow the Chalk outcrop in general, and there are other henges here less well known than Stonehenge, while to the north of Salisbury Plain there is the acme of Bronze Age construction, at Avebury. Silbury Hill, nearby, 'the largest man-made mound in Europe', should have yielded a secret burial chamber but it has determinedly refused to do so. Good luck to it: it is as well to have a few unsolved enigmas. There is no such secret to Avebury. It speaks clearly of being a centre for human activities of all kinds, large enough to enclose the present-day village within its walls, graced with long avenues leading to it lined with standing stones lest you doubt its importance. The rings of great stones in the centre of Avebury still have authority. Here, in unshaped stone, is encapsulated just the feeling that

Henry Moore sought to convey with his 'figures in a landscape'. The stones are sarsens ('greywethers', meaning grey sheep, in local dialect), which also supply the 'uprights' of Stonehenge, and about which more will be said later. How can one doubt the importance of the Chalklands to the early human history of our islands when one sees the prolific evidence of early occupation from which routes spread out far across the British Isles?

Wide, undulating Chalklands are typical of East Anglia, especially in the Gog Magog Hills. Were it not for the general flatness of the area around it is unlikely that the Gog Magogs would have qualified for the name 'hills' at all! At the northern end of its outcrop at Hunstanton, in Norfolk, the Chalk presents a very different appearance: it is red. It forms a distinctive band in the low cliffs: you can collect red brachiopod fossils from it quite readily. Iron compounds produce the red colour. It is a most remarkable departure from the usual white chalk. There are others for the connoisseur of fine differences: the Lower Chalk is grey to the seasoned eye; at places near its western edge in Devon near the bibulously named village of Beer greenish Chalk is coloured by grains of the mineral 'glauconite'.

The Chalklands produce wide rides. The racing community use the open spaces of the Lambourn Downs (Berkshire) and Newmarket (Cambs) to exercise their thoroughbreds. One of the minor contributions of Chalk to landscape is miles of post-and-rail fencing and the neat-and-white yards of stud and stable.

Chalk is not a good building stone. Most of it is simply too soft, although it can at least be cut easily. Some harder beds have been rather optimistically named Totternhoe Stone and Melbourne Rock, cropping out near the East Anglian villages carrying the same names, and these have been quite widely used in local buildings. The general term for chalk building stone is 'clunch' – a rather unfortunate term, resembling the noise one imagines a wall makes when it falls down. However, the Chalk outcrop is so wide in some places that clunch has been employed in churches and cottages; importing stone from elsewhere was too expensive or simply not practicable. Its white colour is not unattractive, and the stone is usually cut into blocks a foot or so square, but it weathers badly; bits flake off it, especially if it gets damp. For older cottages, or small manors, a more usual alternative was timber frame building, with infilling by means of wattle and daub. These buildings are extremely durable. Oak, if kept dry, appears to be virtually immortal. So the Chalk gives us – indirectly – some of the most venerable clusters of buildings in the country. This is particularly the case in areas where early affluence was followed by more modest circumstances militating against extensive rebuilding. Large chunks of Suffolk and Norfolk countryside are almost all timber-frame buildings constructed to a high standard in the great days of the wool trade, the beams being typically plastered over and provided with further plaster decoration called pargetting. 'Pargeter' is still an East Anglian surname, and for all I know there may be pargeters still working, although I have never seen any recent examples of plaster flowers and fruits with the exuberance of the Elizabethan ones. Another consequence of building on the Chalk was a comparable rarity of roofing 'slates', and the Chalklands are accordingly one of the great areas for thatched cottages. Thatch features on American tourist brochures as a prime ingredient of quintessential quaint England: 'England, the country where the houses still wear wigs'. With such nookery-dellery stuff around it is salutary to remember that thatch was used because it worked well using local materials: the Chalk provided no other. This applies to other areas of the country, such as Tertiary East Anglia, with equal force. Reeds (*Phragmites*) were used for thatch that was remarkably durable, but reeds require stretches of shallow, still water in which to grow, and many parts of the Chalklands (not least Salisbury Plain) lack this habitat. In these areas long straw was the usual material for thatching: everything from the wheatfield had a use. This use relied on cereals having long stalks, something which may be helpful for thatching but made the cereals more likely to blow down. Hence plant breeders produced sensibly short-stemmed cereals, which made thatching straw more of a special commodity. Nonetheless, a smart thatch is a particularly handsome roof, well insulated, and its overhang protects cob or wattle and daub from getting wet, which causes rot. Reed thatch may last for twenty years before it has to be renewed. As a child, I even remember seeing thatched walls. The only disadvantage is probably the fire hazard (and the noise made by starlings which are almost impossible to keep out). Much thatch went when roofing tiles became cheaper in the last century. You can readily spot these 're-roofed thatches' by the steep pitches they inherited. There is a final irony to the story of Chalkland thatch. The short straw is almost useless, and some farmers choose merely to burn it, which returns its meagre minerals to the soil. Sparks from such straw-burning are one of the important causes of fire in thatched cottages – and when they burn, they burn to the ground. So what can no longer thatch them, now serves to destroy them.

But the most typical Chalkland buildings are of flint, or to be more accurate, brick and flint. Alec Clifton-Taylor, a normally faultless arbiter of taste in building-stones, is a little disparaging about this combination, but it is a happy marriage which lends character to many otherwise architecturally undistinguished villages. Flint is a recalcitrant building-stone, hard, difficult to shape, with rotten bits if there are fossils in it – but it is tough. The flint can be restrained and guided by courses of bricks which take on the difficult corners or door surrounds. So the result is panels of mottled flints set in mortar surrounded by frames, usually of mellow red brick. Such buildings date from after the appearance of cheap tiling, and so have red roofs rather than thatch. Presumably the bricks were the expensive material, and the flints were local ones. They vary in quality, and in the way they are treated. Most are at least crudely 'knapped', which means one of their faces is broken approximately flat to face outwards: this broken side is usually the darker, fresh flint, and the contrast between the fresh face and the rest of the flint is what gives the stone panels their pleasing mottled look. Near the sea, flint cobbles from the beach are sometimes used whole, selected to match for size and shape as if they were hen's eggs. Sometimes a more innovative builder will make interesting patterns with them. The use of such pebbles extends outside the area of Chalk outcrop because the flint survives into younger formations, with different outcrop areas; and flints are found miles away from the Chalk to produce those pebbly beaches that are such agony to walk across with bare feet into the sea. In Suffolk, flint walls are common in areas underlain by Pliocene rocks. Hence beware lest too simple a reading of local building-stones leads to false conclusions about local geology. I have seen some wonderful 'pot pourri' buildings in East Anglia – even churches – where the builder has obviously been desperate for anything reasonably hard . . . Speckled walls splattered together with bits of clunch and flint and odd exotics from the local beach, even glass.

The most elevated contribution of flint to architecture is to be seen, as one might expect, in churches. The round towered churches are a speciality of Norfolk and Suffolk, though scattered throughout the land, as shown on a map produced by the Round Towered Churches Society, surely the most esoterically appealing of societies. The round towers are usually constructed of roughish flints and attached to small, but charming naves, the whole set in a churchyard of just the right size. Towers make one think of defence, and some of them originally may have had that function in times of trouble, but given the shortage of building materials where they are commonest, no doubt they got their

Above *Pargetting in East Anglia. The decorative use of plasterwork makes a virtue out of the scarcity of good, natural building stones.*

Left *Round towered church, another East Anglian speciality. The towers were usually constructed from flints, of which they make the most economical use.*

shape as a way of, well, cutting corners. If you wish to examine one without going to Suffolk, there is a portrait of the local round-towered church on the label of Bruisyard English wines, one of several good wineries in round-towered church country. However, one could not claim that the treatment of flint was masterly in these delightful little buildings. To see the acme of flint working one has to go to grander churches, perpendicular in style most of them. This is flushwork, the decorative use of flint in small panels, set among contrasting stonework, the flint being skilfully knapped to give a flat outer surface flush with surrounding stone. The flint is carefully chosen, too, to be black and uniform – the same flint that was used by neolithic workmen to make arrowheads. Chequerboard patterns are commonest, black squares of flint and whitish squares of stone. But there are lozenges, and arches, and circles, and all manner of decorative motifs: quatrefoils and cinquefoils, and fleur-de-lys and the emblems of local grandees. Even humbler churches will have flushwork over their porches. The grand ones, like those in Southwold and Blythburgh (Suffolk), have it as well on the tower, along the sides, anywhere it can be seen to advantage. Although these examples are from East Anglia, where the wool trade once easily paid for such extravagances, there is flushwork to be seen over most of the Chalk outcrop and beyond. Thus it is that the skeletons of obscure radiolarians and sponges that lived while the Chalk was being deposited have donated much of their character to dozens of English churches.

Flint, as has been said, does not weather and decay, as does limestone. On the East Anglian coast flint-faced churches have tumbled into the sea as the coast has eroded. Once more they have contributed to the sediment, in the process being recycled yet again. Bruised perhaps, rounded by knocking against one another, these flints will yet survive in future rock.

An extraordinary property of flint is that it fractures like glass, and yet is harder than iron. Break one flint by striking it sharply against another and you will see that the broken surface shows what appear to be ripples. Because these undulations run parallel along a gently curved profile they are similar to the ribs on a clam, and the fracture has been accordingly described as conchoidal – 'as on a conch'. Most minerals and rocks do not break like this. They have preferential planes of weakness that cause them to break along certain lines or perhaps into cubes; or else they crumble irregularly. No matter how hard they may be, they are of no interest in fashioning cutting edges – for as soon as pressure is placed upon them they will tend to fracture further. If you now strike the flint so that a second fracture plane intersects the first you will find that there is a jagged edge between them – and almost certainly it will be a sharp edge. You could easily cut your finger upon it. This is the principle of flint knapping: skilfully breaking a lump of flint to shape.

A well-knapped flint is not a primitive approximation to a tool: it is a perfectly designed implement. But it is not an easy matter to turn a flint into such an implement. It requires practice. The discovery of the special properties of flint must surely have been one of the great technological breakthroughs in the history of Mankind, yet it is probably the least celebrated. The implications of the discovery of the peculiar qualities of flint are more profound than are those associated with the wheel. So many of the functions we associate with communal activity – the very things that make us human – depend on the use of cutting tools. It is difficult to imagine Humankind without them: the capture and preparation of food; skinning and dressing animal hides to make clothes; the shaping of wood for housing; the manufacture of arrows and the like for hunting; and weapons for war and rivalry.

Flint and fire were closely associated: the first fireplaces are close to early well-worked flints. Fire is a knowledge so fundamental that, if we can believe Claude Levi-Strauss, it is built into the very foundation of the human psyche. Fire and flint: two discoveries which established our species. Fire is associated with the Prometheus myth – and it is surprising that there is no comparable myth for the discovery of striking stones to gain a cutting edge – unless it lies so deeply buried in the collective unconscious as to belong to the very inception of our humanity.

Curiously, it is rather easy to understand how early Man could readily appreciate the potentiality of fire. Bush fires happen naturally. It is easy to conceive of a human tasting a cooked animal and finding it good. How much harder it is to imagine the chance discovery of the knappability of flint! Maybe this was the first discovery made by an individual of genius in human history. But other anthropologists would doubtless claim that as the human species spread around the world from its African origins the potentiality of flinty rocks was discovered many times.

There are a few other rocks which have the same qualities as flint, and these have been exploited in the same way. Obsidian is a glassy volcanic rock, black or green and as shiny as a polished finger nail. Obsidian has been used as a tool wherever it occurs – by the Inuit, for example. What flint and obsidian have in common is impalpably fine or hidden crystallinity. Obsidian cooled from a liquid volcanic lava so rapidly that its crystalline structure could not be established: it is a true glass. Flint, on the other hand, originated as a chemical precipitate of silica within a sediment, but its lack of a defined crystalline structure is comparable. Similar, and equally useful flinty rocks occur in other sedimentary formations, and are known by the generic term 'chert'.

The men who knew nothing of crystal structure unerringly selected flint because of its conchoidal fracture, and produced durable cutting edges by carefully directing its breakage. They manufactured tools in great numbers. Sites with good flint became meeting-places, and soon, perhaps, market-places where the goods from one tribe were exchanged with those from another. Marriages could be arranged, deals struck, arguments picked or resolved.

The centres of manufacture of stone tools are described as industries, and if industry is characteristic of civilisation then flint working may be one of the ingredients that brought humans together, and paved the way for society to develop further.

As in most industries, the technology improved with time. Early tools often look little more than natural shapes improved by striking. They sit within the fist fit for dull bludgeoning rather than for delicate paring. But later tools have flaked edges, finely and exquisitely wrought. Here are scrapers and knives and arrowheads and ritual tools each crafted for a special purpose in a special way. The flint demanded for high craft of this quality needs to be flawless. The best kind is black and dense and free from cracks. Norfolk flints from the upper part of the Chalk around Grimes Graves are as white as a freshly dusted loaf on the outside, black as bitumen within. Like truffles grubbed from the subsoil they must have been a precious commodity.

Some anthropologists have sought to learn the craft of flint knapping. M. le Professeur Bordes learned by trial and error to reproduce good stone implements; he could manufacture a serviceable knife in an hour. He deduced that the knapper began by breaking off large flakes from the outside of the flint, to reveal the shape of the tool in embryo. Finer and finer flakes were knocked away using flint adzes, until the cutting edge itself was finished off in imbricate style with a delicate array of fine chippings. The tool was discovered within the rock in the same manner as a sculptor finds the statue within a block of marble. And just as what remains behind in the studio are the chips of marble rather than the sculpture itself, so what often dominates a flint workings are the discarded flakes. Naturally, the useful, worked flint would be taken away for use. But there will also be broken tools, or ones which, nearly finished, developed a flaw. Even the finest flints occasionally have hair-cracks which only reveal themselves at the end of the flaking process. Flakes are not uncommon in chalky areas, and may be abundant if you come across a factory. If you are lucky enough to find a perfect arrowhead it is more likely that it was lost, and the original owner probably cursed when he found it missing as vigorously as we do when we misplace the car keys yet again.

The most advanced stone-workers finished off their tools with fine polishing, to produce axe heads of exquisite proportions; this technique particularly suited rocks softer than flint, such as basalt.

The different industries have been styled with different names, which have passed into common usage: palaeolithic, mesolithic and neolithic – or old, mid and new Stone Age. This is based on the sequence of tools found in western Europe. The Palaeolithic industries were about 400,000 years before the present; 10,000 years ago Mesolithic tools appeared, while the Neolithic industries began about 5000 years ago. It may well not apply worldwide other than in the most general way. There were, until recently, many tribes in the world who were still commonly described as being at a stone-age 'level of development' – most aboriginal tribes in Australia and New Guinea, for example. We would clearly be in trouble if we attempted to correlate societies between Britain and Australasia on the basis of tools alone. It is an intriguing question why some societies should have 'stuck' at this level of development, and when social Darwinism was at its peak guesses based on the notion of improvement led to glib and patronising generalisations about the advance of civilised persons (us, naturally) as against the savage. A purist might even argue that the retention of the old descriptive terms, palaeolithic and so on, might keep a pejorative flavour of these bad old days, but it seems that they still have a purely descriptive value. M. Bordes has demonstrated how skilled was the fabrication of some of these tools; experiments have demonstrated how readily they can be used for the job of skinning and preparing hides, in short how well-adapted they were to the tasks for which they were needed. It all depends on what you mean by primitive. With an almost endless supply of flint, the wonder is that we moved on to anything else.

Is it fanciful to imagine that among the flint workers were the first specialist craftsmen? It seems improbable that all members of the tribe were equally talented at doing the task of knapping – some would have had natural dexterity. The same talented people would have been best qualified to appraise the quality of flint itself – the first geologists – and were certainly able to appreciate enough stratigraphy to follow the good beds of flint into the surrounding Chalk formations in order to mine it at its freshest and best. It is true that the very best flint has to be mined. The countless flint pebbles that form the beaches of Suffolk and Kent do not include many first-rate knappable specimens, and imagine the task of discovering which ones they were . . .

If the flint knappers were the first professional technologists they may have also had time on their hands to perfect techniques and devise new innovations. Idleness – not necessity – is the mother of invention. The hunter and gatherer has little time other than for hunting and gathering. A more settled centre allows the elaboration of oral traditions, the nurturing of older, respected citizens for their wisdom, even experiment. Possibly a caste developed from whom innovation was expected. Dull stuff this silicon dioxide, the commonest pebble in southern England, tough, inert, cursed by gardeners for wedging between the tines of any fork, yet flint may be of extraordinary importance in our history.

A Palaeolithic (Acheulian) flint axe from Swanscombe. Experiments with flint made important early technological advances.

The whiteness of the Chalk has left a curious legacy of monuments. Turf is stripped from the Chalk where the soil is thin. On the downs the grass takes a long time to grow again, and the white patch is visible for miles. Designs cut on the grand scale endure if they are cleaned occasionally. White horses are the most popular design, not all of them ancient. The one at Uffington is genuinely old, and it is this one that has given its name to the Vale of the White Horse.

The sight of it catches you unawares as you drive along the road from Wantage to Swindon, and it is so compelling that you should be warned that you might find yourself suddenly on the wrong side of the road. The design is suggested rather than spelled out, a horse reduced to a few bold strokes. Oddly enough, it resembles the kind of 'logos' developed by Mr Wolf-Olins for big corporations over the late 1980s: the phoenix of the Liberal Democrats is one example. Another ancient one, and much ruder, is the Cerne Abbas giant (Dorset), massive phallus and all. One cannot help wondering whether it was inspired by one of those priapic flints . . .

The Chalk outcrop runs from Dorset to Yorkshire. Salisbury Plain forms its flat climax, while long limbs run across the North Downs and the South Downs, and on the Isle of Wight the Chalk forms the spine of the island where it is tipped upwards, and peters out in the Needles. The Chalk draws the structure of South-east England clearly on the map, with the kind of graphic assurance of the Uffington White Horse. It holds the younger rocks of the London and Hampshire basins, as a hand does a bowl. But the North and South Downs are like the frame of a picture allowing us to see into the older rocks and time of the Weald. Off our islands it runs on into France, into Denmark, and beyond. It underlies the Channel, and forms the matrix for the new tunnel, which largely follows one reliable stratum, from which a giant nautiloid fossil is one of the few contributions to palaeontology from this vast undertaking. Not long ago, geologically speaking, it formed the land bridge to France, before the sea level rise at the end of the Pleistocene Ice Age severed the connection, and thereby guaranteed our nationhood.

The Chalk must once have extended even further over older rocks than it does today. It is preserved in Northern Ireland, protected from erosion beneath the early Tertiary Antrim basalt, the products of a volcanic eruption that spread lava over the landscape burying older rocks beneath a thick, black crust. There is still argument over exactly how far the Chalk sea once covered Britain. But nobody disputes that this sea was one of the great inundations of our islands, indeed of the world. We see nothing of the last, great dinosaurs because of it. In western North America wide-jawed tyrannosaurs, and huge vegetarians, and stegosaurs and crested dinosaurs played out the last act of their long history, while our Chalk sea succoured only sea urchins, and clams, and drifting micro-organisms in immoderate profusion. The only dinosaur from the British Chalk is a fragment, which presumably had drifted out to sea. But there are

The Chalk resists erosion more than the formations above it or below it: The Needles, at the western tip of the Isle of Wight.

other reptiles found as fossils: sea-going ones that pursued fishes and swimming molluscs like ammonites and belemnites, comprising several kinds of plesiosaurs and mosasaurs. Plesiosaurs had a long neck that one can visualise twisting after fast-moving prey, while mosasaurs are more like gigantic, swimming lizards. Very rarely, a flying reptile, a pterosaur, would stray far enough from land to weaken and die in the Chalk sea, its delicate bones faithfully preserved in the soft, white, limy ooze. No doubt sharks crunched many more into pieces before they could become entombed. The sea was very warm. The depth of the Chalk sea has been much argued about – furthermore it varied from one time to another. Even now if you challenge Chalk geologists with the question: 'How deep?' you will not receive the same answer from them all. But most of the Chalk probably accumulated at a depth of about 200–300 m. This may not seem to be that deep, but imagine much of the country inundated to the depth of several St Paul's cathedrals by a warm clear sea, unsullied by much silt from landward sources. It is a unique phase in our varied geological history. The Chalk, as we said at the beginning, is special.

When you wish to understand the genesis of a particular rock type the standard geologist's procedure is the game of hunt the analogue. The world is searched for a similar rock type forming today; if the one from the past looks the same, so the argument goes, then it must have formed under comparable conditions. This is where the Chalk reveals itself as so peculiar, because to find an analogous rock forming today we cannot scout around the continental shelves of Europe, nor

233

even the warm Caribbean or Red Sea. What we need for an analogue for the Chalk is a pure, calcareous ooze composed of the skeletons of foraminiferans and coccoliths. The only deposit forming today which looks plausibly similar is accumulating upon the deep ocean floor, especially in equatorial regions far away from land (remember that there was also little terrestrial input into most of the Chalk). This is known by the euphonious title of the Globigerina Ooze, which one could be forgiven for thinking was the name of a fictional quagmire on the planet Tharg. *Globigerina* is, however, the name of a small and bulbous foraminiferan. The ooze is white, somewhat glutinous, and composed of countless shells (tests) of these single-celled animals, which are only about a millimetre across. The 'forams' live in the sunlit surface of the clear sea and it is the rain of their dead shells through a thousand fathoms or more that builds up the ooze in the deep dark depths, where the resident inhabitants are strange and luminous monsters that seem to be all mouth and stomach. As might be imagined, its rate of accumulation is extremely slow. Early enthusiasts for the similarity between Chalk and Globigerina Ooze led to some absurd suggestions that the Chalk accumulated at oceanic depths. This is just not physically possible, because of the widespread occurrence of Chalk on *continental* areas: the volume of the seas are not sufficient to inundate the continents to oceanic depths. The rest of the Chalk fossils are not those of a deep sea fauna either. So the analogy does not hold in detail; but the different occurrence today of the superficially similar Globigerina Ooze only serves to confirm the peculiarity of this familiar white rock. The truth of the matter is that rocks which we can, today, find in the process of formation at great depths on the ocean floor accumulated widely over the continental shelves of the later Cretaceous.

So we can now try to visualise life on this Chalk sea floor. Over most ·of England (and much further, too) a warm sea swarmed with millions upon millions of tiny plankton. Their dead shells showered through the sea in a continuous drizzle. Sharks lazily thrashed through the water, grabbing what they could. Their shed teeth joined the debris on the sea floor. Crabs scavenged, as crabs will. Occasional errant pterodactyls flapped like reptilian bats overhead. Ammonites and *Nautilus* swam, or crawled over the bottom hunting molluscs and small fish. On the sea floor itself there were areas where little disturbed the white ooze. But elsewhere there were burrows within it. Some of these burrows were u-shaped and occupied by crustaceans – like our living scampi. Others were occupied by the heart urchin *Micraster*, which is the most familiar sea urchin found fossil in the Chalk. These urchins may have burrowed to different depths according to species. Other sea urchins, like those carrying stout club-like spines, without doubt lived on the surface of the white ooze. There were clams there, too. Some were large and reclined upon the sea floor, others had spines which entangled with the sediment. Starfish may have fed on clams, as they do today. There were probably more snails than you find fossils, because their shells require slightly unusual conditions to be preserved. Other kinds of organisms, such as sea mats, took the chance to encrust shells lying loose on the sediment – dead ones, or living ones if they could settle successfully. Brachiopods attached themselves to whatever they could. There were special sea lilies (crinoids) that had lost the anchoring stems of their ancestors and now sculled by pulsing their arms, beautiful and plumose, as light in the water as thistledown in air. There were the ancestors of squid and octopus, too, some of which leave behind cigar-shaped 'guards' known as belemnites. Thus in equable warmth the Mesozoic ended in Britain, and created a rock that stamps its character on much of our scenery and not a little of our history.

For just as the Chalk outcrop defines the essentials of the structure of Southeast England, so its comparative elevation – and the ease with which it could be cleared – made it one of the natural homes for the early farms of the Iron Age, and a favoured site for hillforts with a view. Monuments and tombs cluster there. Tracks as ancient as the Ridgeway, and others doubtless lost to the archaeologist in obscurity, sought out the Chalk for straightness and what we might now regard as exposure. Those roads constructed in the security of modern nationhood skulk in valleys; Roman ones were confidently straight and often imperiously ignored geology; but early routes followed uplands. The same tracks were followed later by drovers, and today are honoured by walkers. Imagine the furore if the same principles were followed today. Motorways would boldly carve along the summit of the Chilterns or the South Downs! Discretion and concealment would be considered a major disadvantage. Local pressure groups would apply for a deviation of the route through their own particular village. Such is the difference in the ideal of communications produced by our age, when appearance is paramount, and threat of attack scarcely a consideration.

As we have seen, the distinctive qualities of flint as a material that could be fractured in such a way as to give a sharp cutting edge influenced the course of development of our earliest communities. The domestication of the South-east

Silbury Hill, Wiltshire. One of the most mysterious of archaeological sites dotting the chalklands.

was fostered by the Chalk; hunted upon, cleared and cultivated, the outcrop interconnected as if Nature had sketched out the routes that she intended us to take in the very fabric of the land. The use made of that land in late prehistory and historical times is a complicated story, described in masterly detail by Oliver Rackham, who has shown how what we see today is the legacy of a past as richly complex as the network of fields, villages, woods and copses itself. This is not our story. But it is worth saying that the wildwood disappeared long ago, and as research proceeds it seems as if man's imprint upon the landscape roots ever further back into prehistory. Even named stretches of woodland may have a continuous history of a thousand years of cultivation, or more. The notion of patches of Roman civilisation amid stretches of primeval forest, maintained by roads and iron discipline, is so much hokum. Iron age cultivation had already spread far, and the Romans found a cultivated land. Fossil pollens from lake sediments put some objectivity into speculations about early agriculture and what happened before it. On this evidence, the wildwood of 4500 years ago was, in the South-east, typified by a variety of lime (prye) woodland which is hard to find in more than small patches today. All has changed, and changed repeatedly. The beechwoods of which I am so fond are not primeval, even though so typical of the Chalk. So what I describe is the land as it is now – at its best in harmony with the geology. It is true that the historical legacy of the Chalklands remains in its plants, and buildings, and streams, and patterns of cultivation. It is only in the last century that our manufacturing developments have sought to impose their own identity on the landscape in defiance of geology.

High Wycombe (Buckinghamshire) might serve as one example of a town growing beyond its natural sympathy with the geology. Somewhere in its centre there lurks a perfectly respectable old place, with a decent church and some old houses. This area is tucked away in the valley, where runs a chalk stream. The furniture manufacturing industry became established there, probably growing naturally from the presence in the surrounding hills of many Chalk beechwoods. Beech makes excellent seats and well-turned legs. At first the town spread discretely along the valley, but then, as the industry prospered it started growing up the hillsides. Because the landscape around High Wycombe is classic Chalk downland these hillsides are rather steep and mostly bare: they command a view. Equally they can be seen for miles. Mass production factories producing sofas proliferated along the valley: some had chimneys which conveyed the smoke up the hillsides as far as the houses. Some of the houses that spread over the downs were built in red bricks that, from a distance, resemble nothing so much as an angry rash upon a pallid face. In any case the houses spread in ranks that have no relationship to the grain of the land, which is why they convey an air of temporary occupation. Like houses made from cards, you feel that a small shake would disturb them, and send them sliding down the hillsides to accumulate in a welter of slabs at the base of the slope. Finally, the M40 has been cut across the valley so that from its viaduct one can see a splendid view of the place. On a still day the smoke lurks in the valley like a guilty conscience.

Further west along the same valley the pretty village of West Wycombe reminds us what things must have been like, once. The small stately home here is the home of the Dashwood family. The ochre wash with which it is covered refers to classical Italy, no doubt, but sits well enough here in this landscape. Francis Dashwood was a notorious libertine, who carved out caves in the hillside opposite. His Hell Fire Club is alleged to have indulged in orgies and satanism in the caves. It has been possible to visit these caves: as might be expected they were cold and clammy, with the walls sweating films of moisture. Rooms opened off the slippery corridors. There was a curious, rank smell; dramatic, but hardly the place to conduct an orgy.

15

The Chalklands:
Beechwoods and Trout Streams

IF WE imagine an idealised chalk garden in front of an idealised brick-and-flint cottage, what should it grow? It should be a light and airy affair, unburdened with those heavy calcifuges that oppress so much of the southern home counties: no rhododendrons, no heather beds (except, perhaps, for *Erica carnea*), few implacable conifers (save yew and juniper), no Camellias. But there are things that rejoice in lime, and many more that tolerate it. There can be an exuberance of roses – even pale pink chalk-loving wild briars, or derivatives of the burnet rose with leaves as neat as babies' hands. There are many clematis that prefer lime: some subtle dark purples, others blowsy and pink and larger than seems natural. There are pinks, which used to be called gillyflowers, a charming name; it would be good to rescue it from redundancy. A cloud of white *Gypsophilum*, like some confection spun from sugar, might contrast with yellow mullein stately as an obelisk. The rare blue of delphiniums should be there in the background, while all the rock roses, *Cistus* and *Helianthemum*, should tumble over paths just as they wish. For annuals there are flaxes – both the scarlet and the clear sky blue, which has recently made a comeback as a crop – and pansies and wallflowers. Gentians and pasque flowers and hellebores will flower late and early, and the leaves of English maple and the flesh-pink fruits of spindleberry colour the autumn. There are some gardeners who seek to defy nature by growing lime loathers on chalk – in special pots, or by watering with compounds that neutralise its effect. This seems an odd endeavour. Why not enjoy this land for what it offers and grow what prospers?

What does Chalk give to the landscape, and what makes its woods, its streams, its flora, and its *feel* so distinctive?

Beechwoods define the chalk ridges. Through the Chilterns and along the North Downs there are dense patches of beech woodland, and even among the ranches of Wiltshire there are hangars of beech trees which stand out the more starkly because they project like mohican haircuts among the bald contours of the hills around. Sparser circlets of beech surround earthworks – hillforts and camps, as at Old Sarum. From the air the Chilterns can look all beech, but on the ground patches of arable and grazing land intervene, to make for a patchy, varied countryside. There are other trees, of course, especially ash in groves on shallow soils, but the pure beech stand is particular to the Chalk. They thrive in its good drainage. The roots fan out in the shallow soil: from an upturned tree you can see the roots splayed out to clutch the ground like the hands of an upended acrobat. The aged, forest beech is a broad-boled, gnarled and massive tree, which may have been pollarded once. In Burnham Beeches in Buckinghamshire some of these monsters remain, and there are a few in Savernake Forest, in Wiltshire. Carbuncular and grotesque, these ancient beeches are climbed by all small children. Not so the slim trees of chalkland groves that rise unbranched to crowns twenty to thirty metres above the ground. In the manner of a tropical rain forest they are all smooth trunk and canopy, and nothing much will grow beneath. The beech leaves and mast even seem to discourage brambles. In spring, beech leaves unfurl, a delightful pale green, fringed with hairs as delicate as eyelashes, and then there is no greater pleasure than walking through beechwoods untroubled by undergrowth in the clear light between slightly silvered, unfurrowed trunks. Later the leaves darken, and there are some forests where you almost need a flashlight to see your way in high summer. This lack of light, and the inhospitable compost formed by undiluted beech leaves, accounts for the rarity of vegetation beneath the trees. Yet there are some specialist plants that seek out this site. Some flower precociously, to take advantage of the light early in the year. The spurge laurel, *Daphne laureola*, is one of these. In March its modest green flowers give off a delicious perfume while light abounds and moths and insects can pollinate it. In summer it skulks unnoticed with leathery and drooping leaves. Then there are those peculiar plants that have abandoned chlorophyll and taken up life as saprophytes living on decay itself, or as parasites on roots. They lack green altogether, progressing directly from root to flower. A fungus helps the saprophytes break down the humus. The brownish flowered Bird's Nest Orchid (*Neottia nidus-avis*) is not uncommon, but the Spurred Coral Root, *Epipogium aphyllum*, is one of Britain's rarest plants. It disappears for years, only to be rediscovered in a new locality, never with more than a few individuals. It has a range from Britain to Japan, but is common nowhere. From its habit of disappearing, and then mysteriously reappearing it is sometimes called the ghost orchid. Another curiosity is the toothwort, *Lathraea squamaria*, a parasite of roots with pallid mauve flowers. In my experience it is commonest in hazel coppices, which are frequent in parts of the Chalklands, though now usually overgrown. The name 'corpse flower' has been applied to this plant, allegedly because it sprung from buried bodies, but it seems plausible that the colour itself was suggestive of rigor mortis. The largest flower in the world belongs to another root parasite, *Rafflesia*, named after Sir Stamford Raffles, who also gave his name to the hotel in Singapore. It springs massively from roots in Sumatra. These flowers truly smell like corpses. It is strange that both our humble toothwort and the spectacular *Rafflesia* have these connotations of death: there is something about the lack of chlorophyll in a flowering plant which goes against the natural order of things.

There are other rare orchids that follow the Chalk outcrop. Some of these are now so rare that their localities are protected. Apparently rarity is an irresistible lure to some people – they will sneak in at night and attempt to dig these plants up: monkey orchid, military orchid, early spider orchid, musk orchid, lizard orchid; all beautiful, all rare. But the bee orchid is much less rare and can give as much pleasure. It seems to like quarries that are reverting to grass; it used to grow in some profusion at Box Hill on the North Downs, and perhaps it still does. It will grow away from chalk, but seems to prefer alkaline soils generally, and is found on other limestones. It is famous for the fluctuations in its abundance. Hundreds can be found in a favourable site one year, and when you bring a friend to see the wonderful sight the following year you are embarrassed to be unable to find more than half a dozen. In spite of its resemblance to a bee it is not one of those orchids which is pollinated by virtue of a male bee mistaking the flower for a female and attempting copulation – thereby releasing a clever pollinating device. Charles Darwin made these mechanisms famous in *The fertilisation of orchids* (1862); apparently the bee orchid prefers self-fertilisation.

The pleasure of lying on short turf examining your first bee orchid will not be forgotten by any naturalist. Strangely exotic for the home counties, it looks like an escapee from a hothouse. In June the chalkiness of the downland soil will be obvious, and the chances are that it will not be long before a protruding flint forces you to shift your position; or perhaps you will sit on the barbed rosettes of the stemless thistles which grow among the orchids. Look closer. A mere breath of a plant nearby is the fine fairy flax (*Linum catharticum*), with tiny white flowers, and certain to be found on any chalk hillside in early summer. Its Latin name shows that its purgative effect belies its delicate appearance. There may be a patch of bright blue milkwort (*Polygalum*). This is one of the few plants which can have quite differently coloured flowers within the same wild species; pink or white varieties are rather common, and once I found one blue with white tips to the petals. Even the grasses on Chalk Downs are special: quaking grass with shaking flowers, or heavy bromes. As with the fairy flax, many plants demand you look closely – chalk turf is like an alpine garden. Some plants are fragrant, like the wild thyme (' 'Tis good for Spitting of Blood, and Convulsions, and for Gripes' according to John Pechey's Herbal, 1694) which always seems to grow on the ant hills made by red ants; or wild marjoram heavy with bees. There is even a butterfly – the chalk hill blue – that relies for the successful growth of its caterpillar on the association of thyme and anthill. So a passing butterfly may be tied to the geology through the subtlest threads of dependence . . . surely there must be other connections still more arcane that we do not yet understand?

To return to the beechwood, there are other inhabitants dependent on both beech and chalk which are invisible for ninety-nine per cent of the time. To see them you have to visit in October after rain. These are the fungi. They grow hidden as fine, white threads of mycelium feeding in the humus for most of the year until autumnal rain encourages their fruit bodies. The toadstool is simply an umbrella raised for the shedding of a million spores. There are dozens of species that live only here. Some grow so closely in association with beech that they cannot be found elsewhere – they have a symbiotic relationship with the roots of this tree alone. The brilliant coloured Russulas, which can be green or orange or carmine, are the most obvious of these, but there are slimy *Cortinarius* species as well, and ceps (which all used to belong to the genus *Boletus*). *Boletus* have sponge where most other mushrooms have gills, and this makes them easy to recognise – moreover there are very few species that are poisonous and many that are delicious. There is a puffball (*Lycoperdon echinatum*) that looks like a rolled-up hedgehog, and very occasionally the magnificent coral fungus (*Hiericium coralloides*), an unmistakable bouquet of creamy spines hanging like some extraordinary terrestrial coral from rotting stumps.

There was a time when I was the only person with a mushroom basket snuffling among the beech mast, for the English have had a terror of any mushroom other than those that are white and grow in greengrocer's shops.

But in France murder has been committed over a basket of chanterelles, and I cannot avoid a twinge of resentment when some of my favourites have been snaffled by someone else. Most hidden of all are the truffles (*Tuber*) – subterranean fungi which have to be discovered and disinterred. The English truffle (*Tuber aestivum*) is a different species from the famous Perigord truffle, which is the most expensive foodstuff there is – more expensive than Beluga caviare. Its aroma is a kind of gastronomic aphrodisiac. But *aestivum* is reported as very good, and there were professional truffle hunters in England until this century. In truth it was trained animals, dogs, or pigs, that did the hunting, sniffing out hidden truffles with sensitive noses; their owners merely removed the booty. There is a special fly that feeds on truffles, and it is occasionally possible to locate a truffle by looking for a crowd of dancing flies over a beechwood 'floor'. I found one once only in the Savernake Forest (Wiltshire); it was too ripe, but still smelled delicious. The appearance of this summer truffle is not appetising – like a black, warty growth. Today the art of training truffle dogs has been lost, but with other former professions enjoying revivals, like weaving with osiers, or trug-making, perhaps it is time for the reappearance of the truffler.

While most fungi of beechwoods on chalk are autumnal there is one vernal species that is sought after as much as the truffle. This is the morel (*Morchella esculenta*). This species looks weird – a brown, wrinkled brain on a white stalk, up to 30 cm or so high. It is not closely related to the cultivated mushroom, but belongs instead with a group of curious and often brightly-coloured cup fungi which also mostly crop in spring. The only places where I have found morels in abundance were all on chalk, and in woodland, although ash rather than beech was the dominant tree. The discovery of a cluster of morels is thrilling, not least because they are rather rare in this country. In spite of their distinctive appearance they are easy to overlook tucked among the wayside foliage and dead leaves. They are reported to favour burned ground, although that is not my experience. In northern France, they even started bush fires late in the year to encourage their appearance the following spring. They make wonderful eating, with a smoky flavour that is found in no other mushroom, nor any other vegetable for that matter. French cookery deals with them in a hundred different ways, but Jane Grigson might be right when she says they are best à la crème. Neither bullying nor bribery would make me reveal my own morel sites.

Beeches have been much planted in the past, and the chalk beechwood is in no sense wildwood. It has been cultivated. The timber of the beech was extensively employed in the furniture trade. Itinerant wood turners took to the woods and made chair legs. They were called bodgers; it is reasonable to infer that their workmanship was not always of the highest standard.

Two chalk natives are evergreens. One is the yew (*Taxus*), but it is not usually encountered in pure stands. Where it is, as on the Chilterns near Watlington (Oxon), or at Mickleham Downs (Surrey), a sinister woodland results, even darker than a beechwood in summer. The roots make fantastical cages where the soil washes out beneath them . . . you would scarcely be surprised to see an Arthur Rackham gnome sewing a leather jerkin in one of these places. Similar darkness is generated beneath box (*Buxus sempervirens*); its specific name translates from Latin as 'always green', and so it is, although with its delicate little leaves it always seems a surprising evergreen. Native box trees are scattered in patches in chalk woodland; these are not the civilised, clipped dwarves of the Elizabethan formal garden, but are straggly and sometimes rather unsightly small trees. Box timber is famously hard, and so valued in the past that wild box trees are not as common as they might be. Box Hill (Surrey) and Boxley (Kent) used to supply the London comb-makers, but I doubt if it is possible to obtain such a comb now, and tough plastics have also made the boxwood ruler obsolete. Box groves are protected by the National Trust at Box Hill, and it will be interesting to see if they spread more widely now that the demand for box timber has fallen.

On the edges of chalk copses, in hedgerows, and in overgrown quarries there is one of the most characteristic lime-lovers of all: the wild clematis, or old man's beard. It is the English liana, with improbably long stems – hardly thick enough to be called trunks – straggling up trees and quarry sides. It is so sensitive to lime that it can be used as the basis for a geological map – even if you cannot see the rocks. The flower is a disappointment when you think of the garden species of clematis, being a small white thing. But the 'beards' that follow are spectacular. Each strand of the 'beard' is attached to a seed, forming round heads. In early autumn it looks as if wild sheep have extravagantly scraped off tufts from their fleece all over the hedgerows.

All kinds of chalk woodland, especially in the home counties, probably owe their continued existence to the pheasant. The arable vastnesses of the rest of the chalklands show what happens when the land is turned to immediate profit. But the undulating woodlands flanking dry valleys are suitable for game birds, and there are pheasants and partridges (rarer than formerly) and red-legged partridges. Compare this with the wastelands of wheat: a mere one species of huge commercial importance. Selective weedkillers have made the former chalk cornfield weeds like Corn Marigold, Corn Buttercup, Venus' Looking Glass, and Pheasant's Eye as rare as protected orchids – more so, because reserves are not set up to conserve them. On the thin, chalky soils, wheat and barley can only grow if fed with nitrogenous fertilisers. These chemicals drain off down the slopes, and the main beneficiaries are the field boundaries. These were once a mass of scabious, knapweed, and interesting parsleys. But all that can cope with hefty doses of nitrogen are nettles, docks and cow parsley, the rankest and least attractive of herbs. In high summer a sense of oppression emanates from these endless swathes of rich, fed green, where once there was variety and colour. It is curious that rich soil reduces variety. The most varied chalk meadows, never ploughed, were those where sheep grazed for hundreds of years. The wool removed nutrients. The sheep cropped closely. Both processes encouraged the chalk flora, but when cows replaced sheep, and fertiliser was added to stimulate growth of the kind of grasses cows preferred, most of the chalk flora simply disappeared. It is too late to mourn its passing: small patches survive. Local Trusts for Nature Conservation have preserved occasional bits of 'managed' meadowland which somehow had escaped the process of change. A good example is at Streatley, above the Thames in Berkshire. But it is more of a pleasure to come across unexpectedly a patch that has survived. I found such a patch in the middle of a golf course where grew yellow-wort and pyramidal orchid and wild mignonette. There one could recapture for a moment something of the variety that the Chalklands have lost. On a hot day the bees outdo the thwack of golf balls.

There is no better example of the connection between geology and woodland and flora than the Chalk. Because so much of its outcrop surrounds London there are more than the usual number of people vigilant for the protection of its good things. Footpaths are defended. Even the verge of a footpath can act as a microscopic reserve, unless it has already become totally swamped with nettles. It is conceivable that the switchback of economics might one day establish sheep as profitable again – would the chalk pasture then return with all its species? The reserves do a wonderful job, but is it right that we should go there to admire chalkland orchids in these reserves as we would animals in the zoo? It seems rather unlikely that seeds of all those lost species survive in the thin soil. Some seeds are designed to survive for years – decades even – until they once again meet the right conditions for germination, but research has shown that this does not apply to most orchid seeds, and there must be others that might be lost forever. Equally, one cannot 'fix' the countryside at a particular stage in its development. If conservation had been invented centuries ago there would no doubt have been lobbyists objecting to the destruction of the wildwood and the interpolation of all those nasty fields. Paradoxically, previous human meddling actually increased the variety of habitats. Cleared woodland, sheep grazing, hedgerows, cornfields, even quarrying, have added more small habitats for more species. Overall, the chalk habitat was enriched; the wildwood may even have been rather monotonous. Variety was maintained in the past as a by-product of perfectly ordinary farming practices.

Springs well up from deep in the Chalk to feed chalk streams. Springs are pure water, clear and cold. When they bubble from the ground it almost looks as if the water is boiling. The water finds its way to the surface along fissures or joint planes within the rock. Several springs serve to feed the stream, and more feed it as it proceeds. Because it is spring-fed, the level of the water does not relate necessarily to the amount of rainfall that has recently fallen: it can take months to find its way through the Chalk. It is even possible to find a stream which deepens suddenly in the middle of a dry period as earlier rain finds its way through the water table. Chalk streams (bournes) are swift-flowing near their sources. Flint pebbles floor them. This is the habitat of the freshwater crayfish – the largest of our crustaceans to live away from the sea. You have to be very still to see them; often you see their sensitive antennae alone projecting from beneath the flints. The look a little like lobsters, perhaps, and they are equally edible, although they are no doubt protected. Its giant Australian relative, the yabbi, is delicious. A little fish that we used to call the 'miller's thumb' is found in the same kind of shallow stream. Never having seen the thumb of a real miller I cannot say whether the fish resembles the model. The streams abound with water crowfoot (*Ranunculus* species), which light up the waterways with white flowers in late spring. These are the subaqueous relatives of the common buttercup, although you would never guess so from the feathery, deep green leaves, which toss like wind-blown hair in the brisk currents. This water is clear, and so it is ideal for the hang-over-the-bridge-and-see-what's-going-on kind of afternoon.

I am not sure which is the perfect chalk stream. Possibly it is the Lambourn, rising on the downs where potential winners are exercised and flowing south-wards to the Kennet along its own valley. I was told once that it has the coldest water in the South of England, one of those claims, like villages that state they are the middle of England, that seem difficult either to substantiate or refute. Spring water that has come from deep in the chalk *is* likely to be cold. Possibly, the Hambleden Brook would have been another contender for the ideal chalk stream, which runs off the Chilterns to the Thames. The village of Hambleden is almost too perfect; it has been used for the setting of almost every Agatha Christie murder mystery where the ideal village is required; also the water meadows have been re-sown once too often, and so lack buttercups and ragged robin. The Winterbourne, south of Amesbury and followed by a wonderful by-road to Salisbury, must be another candidate. Surely, the name Winterbourne must refer to that chalk stream's propensity to flow more at some times than others.

Even as they get deeper chalk streams remain clear. The Test and the Itchen and the middle reaches of the Kennet continue to flow at a decent pace. The southern England streams are hardly polluted – at least by industrial effluent – and this has to do with a fact, which even I cannot claim has wholly to do with geology, other than the alkaline purity of the water. The middle stretches of chalk streams are the ideal habitat for trout. Brown trout, in particular, are considered the acme of sporting fish. The most exclusive club in the world is not White's, nor Boodle's, but The Houghton Club. Membership of this club entitles you to fish some miles of the River Test at its prime near Stockbridge. No lob worms or elderberries allowed here. Only fly fishing is permitted. With successive casts, the 'flies' repeatedly float past above the nose of the quarry, which, at the right time of day, usually dusk, may, just possibly, take a lunge – whereupon the fisherman must strike at exactly the right moment to engage the tiny, hidden hook secreted on the 'fly' into the fish's jaw. It is a decidedly difficult art, and requires deep concentration. Some of the species of fly which are imitated are chalk stream specialities. Dun little things, their imitations have none of the extravagance of the vulgar salmon fly (which seeks not to imitate but only to attract), but one might say that they, too, are part of the chain that connects with the rock beneath, because they are so particular about their water. This choosiness comes about because their larvae live in the stream for up to a year before they become adults, and only pure chalk water offers exactly the right habitat. For the angler, the high time of the year is when the mayfly hatch and dance and lead the briefest of lives to mate before they die; they are juicier than most flies and the fish go mad for them. A river keeper told me that the mayfly is much less common than it used to be. Brown trout, of course, occur elsewhere in rivers other than chalk

Dry valley in the Chalk: Devil's Dyke, Poynings, West Sussex.

streams, but the special qualities of the Itchen and the Test breed special sport, and perhaps we should be grateful that they do, because it seems doubtful that the purity of the water would otherwise have been monitored so carefully for so long. Even so, one suspects that seepage of nitrates may have altered water quality enough to upset the mayfly.

There is one kind of valley in the Chalklands which has no stream running at its floor. Yet these valleys are steep-sided and obviously were, once, cut by the action of streams – and by vigorous ones, too. Such dry valleys often branch off deeper valleys in which a chalk stream still flows. One dry valley leads you through miles of Chiltern country towards the Warburg Nature Reserve, near Henley, a charming little road now following the course where you might have expected a little bourne. There are many around Selborne (Hants), the home of Gilbert White, the natural historian who still delights us with the observations he made in the Chalklands of two hundred years ago. It is curious that so acute an observer did not make more geological observations. His walks must have led him along dry valleys but I am not aware that he speculated on their formation. The vigilant naturalist might deduce that the streams which once cut such valleys have now disappeared downwards into the Chalk itself, following a lowering of the water table. The 'winterbournes' even now appear and disappear. It is likely that the carving of the valleys happened when the climate was much colder than it is now, towards the end of the last 'Ice Age' (about ten thousand years ago). If the ground were permanently frozen at depth as it is in Arctic areas today ('permafrost') erosion could have proceeded much faster, particularly when the streams were fed with melting snow. I have seen such streams rushing energetically in the Arctic summer, even bouncing stones along their beds, and one can readily imagine such streams carving valleys. Even now one can find beds of flints in the floor of dry valleys, the very tools which did the carving. And there is further evidence of the action of ice in the 'combe rock' which often fills the valley sides, a chalky-clayey mixture that was produced by the freeze-thaw cycle of the glacial year.

On some areas marked as Chalk on the geological map, you may find patches of plants that seem most inappropriate: harebells, or gorse, for example, which are usually acid lovers – even lime-hating rhododendrons. It is likely that these interlopers are growing on patches of clay-with-flints, and one cannot leave the Chalk without mentioning this superficial deposit which contrives to countermand most of the generalisations that have been made about what grows on the Chalk. For clay-with-flints is what is left behind after long leaching and solution of the Chalk, all the minerals *other* than lime, and, naturally, the indestructable flints. Clay-with-flints often perches on downland, like the small patches yellow with gorse dotted along the Marlborough Downs. It is yellow-brownish, heavy, sticky stuff, stuck with sharp, hard flints that make it extraordinarily difficult to dig.

The youngest Chalk in Britain is in East Anglia, but nowhere in this country do we see rocks that tell the story of the very end of the Mesozoic, the demise of dinosaurs and ammonites, and the inception of a world more like our own.

16

Tertiary Times

AT THE western end of the Isle of Wight the Chalk is turned steeply upwards. Its outcrop forms an obvious backbone to the island, and because it is tougher stuff than the rocks above it or below it, it makes up steep cliffs, which continue westwards as The Needles, the most celebrated marine stacks in the British Isles. The steep dip of the Chalk here means that its outcrop does not take up a great tract of countryside, as it does when it is horizontally disposed, of which Salisbury Plain is the example. Instead, you can drive across its whole width of outcrop in a minute or two. The fold in the Chalk here is but a tiny wrinkle in comparison with the magnificent contortions of the Caledonides, or the jagged folds of North Devon, but it still comes as something of a surprise among the gentle undulations of the South-east. It used to be described as the last gasp of the great earth movements that threw up the Alps – at root the inevitable consequence of the northward movement of Africa. Nowadays, and perhaps a little more prosaically, the fold is thought of as a superficial ruckle produced in reaction to movements on deep-seated faults, a wrinkle in the icing produced by stressing the cake. Whatever the cause, you can visualise one of the most important events in the history of the Earth thanks to this fold turning the book of rock history upright.

Alum Bay is famous for its coloured sands, which are sold packed in hour glasses and miniature lighthouses, in layers of white, pink and creamy yellow. The sands derive from the Tertiary rocks that overlie the Chalk. Ignore all the commercialisation at Alum Bay, if you can, and walk down to the sea shore – there are a lot of steps – or take the chair lift if the walk up again seems intimidating, and there you can inspect the source of the Alum Bay sands, and see what we have in England to record the passing of the dinosaurs.

The long, clean cliff marking the top of the Chalk lies to the south, running westwards out to the Needles; the steep slope down into the sea approximately marks the bedding surface, and because this is part of the famous fold, everything lying nearer to Alum Bay is younger than the Chalk. At the time the topmost beds of the Chalk were being laid down beneath the sea, there were still dinosaurs roaming the plains of Canada, and ammonites in the sea. But by the time the first of the dark clays that overlie the Chalk were deposited, which you can examine in the corner of Alum Bay, everything had changed, changed utterly. The dinosaurs had been exterminated, likewise the ammonites that had had an even longer tenancy of the Earth than the dinosaurs. Scarcely any kind of animal was left untouched by this change between Cretaceous and Tertiary. What shows here as the change between resistant chalk and soft sands and clays – so soft they form a bay that enlarges with each year, each winter storm – was the change between an archaic world and a world recognisably our own. While Alum Bay sands accumulated in their shades of silver, purple, grey and even green, there were mammals upon the nearby lands that were not our mammals perhaps, but would have been recognisably furry and would have cared for their young, or even preyed upon one another. There would have been birds in the trees, and many of those trees would have had broad leaves, like magnolias or maples, while on the plains a burgeoning carpet of grass was laying the basis for our living ecology. Between Chalk and Alum sands a great meteorite, or maybe several, blasted the Earth, the climate changed several times, the seas became fouled, and reptilian domination passed from the world forever.

To say that scientists do not yet agree on the details of this scenario would be an understatement. What they all agree upon is that a profound event, or events, happened at the Cretaceous-Tertiary boundary, and happened fast. Those who favour the meteorite – which has left its chemical signature imprint upon dozens of rock sections around the world – would like it to be the prime mover, initiator of all other effects. 'Darkness covered the Earth, and gross darkness . . .' caused the death of green plants, and provoked a general cooling, which together destroyed herbivorous dinosaurs and the meat-eaters that preyed upon them in months, if not days. The same effects provoked death in the sea, and the demise of a thousand humbler species. Not so, say other scientists, the climate was already in deterioration, and a cold spell alone would have doomed dinosaurs to death. An additional cause, say another group, was changes that happened in the sea, with the shift of sea level stimulating widespread changes on land, in the latter case by oxygen-starved waters becoming ubiquitous enough to poison all but the hardiest. They point to evidence which shows that the dinosaurs were already declining by the late Cretaceous, and claim all that was needed was a final *coup de grâce*. Feeble-minded prevaricators like myself would probably like to claim that the meteorite was undeniable, and the other factors likely, so that the cause of the dinosaur demise was a disastrous combination of all these proximal causes, the global equivalent of the day grandmother's Ming vase got broken, the car was dented, and you broke your arm on the way back from the garage.

At Alum Bay signs of before and after the great 'event' are written clearly in the rocks, and there is no better place in Britain to think about how momentous have been the changes in the geological past that have led the Earth to where it is now. Can one in a thousand of those who play the fruit machines at the clifftop realise that the stamp of the most disastrous combination of mischance in the history of the Earth lies exposed in the cliffs beneath? The tumblers of chance clicked and the Earth changed forever. To be completely accurate, the fateful moments themselves are not recorded in the cliffs in Alum Bay, because of a break in sedimentation there at the critical level. The topmost parts of the Chalk had been eroded away before the deposition of the early Tertiary. To place one's finger on the moment itself, it is necessary to travel to Gubbio in Italy, or to Cantabria, or Tunisia, or the western USA. Nonetheless, as you face the cliffs in the Isle of Wight, to your left lies the future, and to your right the past, all turned upright so that the passage of time flashes before your eyes like the flipped pages of a comic book.

In the face of this drama one must not forget the alum, which after all gives the bay its name. This is a sulphate of potassium and aluminium that flushes out of some of the darker shales and clays of the Tertiary. It forms a yellowish, fluffy efflorescence near the shore in several places. These rocks are a regular hotch potch of different colours and textures, with irregular lenticles of conglomerate or gravel, as well as sands and clays, and there is even a band of coal. All the rocks are soft. There are very few horizons younger than this in Britain where there has been time enough to harden the rocks enough to trouble a geological hammer. And in the places where fossils occur they can often be plucked out, like nuts from a cake. The unwary collector might confuse a genuine fossil with oyster shells left by a Victorian labourer, or a whelk shell dead for a decade or two. Some of the real fossil shells can even be close relatives, biologically speaking, of such historical debris. The fossils have usually been *indurated* to some extent, so that the microscopic pores in the shell have become filled with a mineral such as calcite, which makes them suspiciously dense. Often, also, they have taken on the colour of the enclosing rock, yellowish from an iron-rich sand, or greyish from a clay. To be sure to avoid confusion, fossils are best collected directly from the cliffs.

The dip of the beds soon flattens out northwards away from Alum Bay; within a few miles the rocks are nearly horizontal, and so they remain over much of the Tertiary outcrop. There is not much natural exposure apart from in sea cliffs and along river banks. But because much of the Tertiary comprises sands and gravels there have been numerous pits opened up in them; these have been worked to supply hardcore, and ballast and the makings of concrete. For the Tertiary underlies the South-east of boom development and speculative estates, of DIY superstores and mock Tudor. There is scarcely an arterial road that has not been widened, nor a piece of waste ground that has not been turned to profit. The subsurface is probed and pierced and augured and pile-driven more than the rest of Britain put together. A lot is known about its geology, although much of it is more deeply interred beneath human construction than are Precambrian schists beneath the blanket bogs of the floe country.

The Tertiary rocks cover two great areas: the Hampshire Basin and the London Basin. Alum Bay lies at the south-western edge of these. At the present day older rocks of The Weald and the Hampshire Downs separate the two basins as a consequence of gentle folding which affected the whole of the South-east, but the two basins would have been contiguous when the rocks were laid down. Tertiary rocks carry on into France and Belgium, and the Paris Basin, all of which combine together with the British deposits to comprise one huge former geographic entity. The interpolation of the Solent and the Channel are recent digressions which disguise this deeper reality. Some formations can be traced over the whole area. In Alum the width of their outcrop is extraordinarily narrow because of the way they have been tipped vertically, but elsewhere a single formation may cover hundreds of square miles.

The best localities to examine the rocks and collect fossils are probably at Barton on Sea and Whitecliff Bay (Hampshire). As with so much of the South coast the clifftop countryside has long been subsumed under retirement homes called Holmleigh and Bay View. I have never actually seen a *Mon Repos* or Dunroamin, but I have always cherished such confections, and plan to call my own retirement villa 'Dungeologisin'. Neat and wholesome thirties-style houses predominate at Barton, but there is less affluence than there once was. The cliffs have a life of their own. The clays often slump and slide, or are attacked by particularly vicious storms. This has a bad side and a good side. The bad side is that it is often difficult to find a bed or an horizon that a guidebook tells you should be readily recognisable. What may have been prominent one year may be concealed the year after next. The good side is that new things may be revealed by a fresh slump or storm. Such localities are never collected out, and once in a while something exciting turns up. Birds, for example, are very delicate and rare fossils; new species have been discovered on the Hampshire coast. This is one of those places where it is best to live locally if you are going to be a palaeontologist – to be first on the scene after a storm, before things begin to disintegrate. Some of the collections made by local enthusiasts are now ornaments of the national collections in the Natural History Museum. The interests of the geology and the local council have clashed more than once; the latter would like to encase the whole cliff in erosion-proof concrete, and it is perhaps fortunate that this substance does not exist. One of the formations on the Hampshire coast, the Barton Formation (or Barton Clay as it used to be called) of Eocene age, is extraordinarily rich in fossils, especially of extinct, subtropical marine snails. These total six hundred species! They are easy to find, too, and some, like the large ribbed conch *Athleta*, are also ornamental. The tiny species are even more interesting. They can be washed out of the clay at home, where an amazing array of shapes and ornament can be released after only half an hour's work. It is only sad that original colour is not preserved: Barton Clay fossils are dull-coloured, or with a pinkish tinge, but were no doubt originally speckled and blotched. There are even turtles and crocodiles to find from time to time.

What we owe to the Tertiary in the Hampshire Basin is the New Forest. In spite of the queues of cars that clog its main arteries in the summer, the Forest manages to preserve something of wildness – even of wilderness. Verderers still govern it. A little pannage is still practised, whereby those living in the Forest have rights to fatten their pigs on plump acorns fallen from ancient oaks. The well-known ponies are another sort of crop.

There are families who have been there for more generations than they can count, and this is the kind of loyalty to place that one also finds in the Forest of Dean. To judge from the rarity of the grander kind of ancient cottage within the Forest it must have been an area of mean subsistence in the past. Cottages built

The unremarkable cliffs at Barton, Hants, have yielded a prodigious variety of Tertiary fossils: snails, clams and vertebrates.

before the nineteenth century tend to be walled with cob, and deeply thatched, with upper storey windows peeping out – all picturesque now, but probably very uncomfortable when there were earth floors and scarcely a stick of furniture.

The New Forest, like all forests of ancient origin, was a royal preserve for game, and especially deer. The modern use of 'forest' for extensive commercial plantations of trees is a misleading transmutation. Not much of the New Forest, only about a fifth, is unbroken trees. It has great tracts of true heathland, and this alone would make it of outstanding importance, particularly as heathland elsewhere is disappearing fast. This is where ling, gorse, adders and specialist butterflies thrive, and where human devotees hunt interesting warblers. Where drainage is poorer, valley mires comprise a wetland habitat that is now rare or fragmented elsewhere in Europe, and this is where Sphagnum moss grows and peat accumulates, as it does in Caledonian moorland. Alder trees and sallow shrubs grow at their centre, together with great clumps of tussock sedge (*Carex paniculata*), which lends these mires a completely different appearance to the Caledonian wetlands. The fen-woodland flora is a mass of delicately beautiful plants – yellow and purple loosestrife, attractive water mints, spearwort buttercups, all grown through with clambering marsh bedstraw. In places, bog myrtle thrives – a plant more usually associated with Ireland or Scotland – which has leaves that yield an exquisite fragrance if they are rubbed between thumb and forefinger. The only other place I have smelled a comparable smell was in the Australian outback, when a handful of eucalypt leaves were thrown on the embers of the fire.

Grassland areas are also rich in plant species: many grassy patches are appropriately described as 'lawns', because they have a short and even sward, which has been produced and maintained by centuries of grazing. There are species of plants here that grow wild nowhere else in Britain, most spectacular of which is *Gladiolus illyricus*, the likely ancestor of the more lurid cultivated 'glad'. It is common in Mediterranean regions, and still somehow, when you find it near Lyndhurst, looks irretrievably exotic, native though it may be.

Grazed, open ancient woodland completes the inventory of major New Forest habitats. Some of the trees seem of incomparable antiquity: massive oaks with dead 'stagshorn' branches, or ancient beech pollards long since grown into giant trees. However, even the oldest oak for which tree rings have been counted scarcely topped four hundred years, which is not a great age for *Quercus robur*. Trees, even oaks, do not live long on the poor gravelly soils of the New Forest. However, oaks were much harvested to supply navy construction in the seventeenth and eighteenth centuries. The woodland is home to the only British cicada, *Cicadetta montana*, and even here it is confined to one small area. The open woodland has the richest lichen flora of any lowland wood in western Europe. But the carcasses of dead and dying trees that have been allowed to lie in the Forest are another special habitat that is becoming rare elsewhere. Many kinds of beetles are found nowhere else. Hornets are perhaps the least welcome inhabitants of holes in standing dead wood. There are fungi that break such dead trees down over decades, and these in turn have their own succession of different beetle species that feed upon them. Woodpeckers peck the trunks for grubs. The Forest has, in places at least, avoided the mania for 'tidying-up'. May this patchy, various place remain just as it is – and, not least, avoid getting fussed over too much. Excess of zeal has frequently proved more damaging than benign neglect.

The poor soil of much of the New Forest is the consequence of the Tertiary rocks that underlie it – it is doubtful that the Forest would have survived in so complete a state since the thirteenth century if the rocks had everywhere been capable of breaking down into a rich and productive loam. The richest soils are found on the clays and marls of the south of the Forest, where human habitation also has the longest history.

The landscape itself is largely the product of several gravel terraces which make the higher, flat ground. Although the high point is only at 128 metres, clear panoramas make it feel much higher. The terraces are underlain by more or less coarse gravel – yet another manifestation of the durability of flint – but this deposit is superficial, lying on top of various of the Tertiary formations, and hence is evidently fairly recent, in geological terms. It is usually only a metre or two thick. Nonetheless this thickness is enough to produce the poor soil and good drainage which is needed to foster heathland. There has been much debate about how the 'plateau gravels' were formed. The explanation favoured at the moment, which was evolved by David Keen in 1980, is that they were river gravels deposited by the 'River Solent' in the cold of the Pleistocene Ice Age. As sea level fell, the river gradually moved south-eastwards towards where the Solent is today, in the process creating flat, gravelly plains. The gravels were the product of near flooding of meltwater each spring, as braided streams choked on the abundant runoff.

The underlying Tertiary rocks are oldest in the north of the Forest, where they are also poorest in nutrients: here some of the beds exposed in Alum Bay crop out again – the Reading Beds and London Clay. Bagshot Sands a little further to the south underlie extremely poor soil, which is good for nothing in crop cultivation, and results in heath and bog alone, as at Cranesmoor. Fossil pollens from the bogs tell the story that the spread of poor heath may have been comparatively recent, perhaps even following human clearance of woodlands which formerly covered the area. Without the canopy of trees, nutrients were leached out from the thin soils, and once this happened only bog and heath could flourish. Heath fires – which still happen regularly in dry summers – discouraged the return of broadleaved trees, and encouraged plants like gorse or ling which could regenerate readily after burning. If this story is true, it makes an interesting comparison with what is happening today in tropical rain forests, where most of the nutrients are also tied up within the trees, and the soil is basically poor. Clearance, and the subsequent planting of conventional crops soon exhausts the soil; but by then the ecosystem has been changed and cannot go back again to the same forest which grew before. The difference between the New Forest and Amazonia should be thought about before we westerners start getting in a moral dudgeon. But from a narrow, utilitarian view it can be said that the heaths, bogs, and their flora in Britain actually *add* to the variety of habitats and species, whereas the Amazonian case is a notorious impoverishment of the richest assemblage of species anywhere. It is interesting that an ancient example of similar processes may lie within our own comfortable little kingdom, and an irony that its aftermath may be protected by conservationists.

One of the fossils found in the lower Tertiary rocks of the Hampshire Basin is a discus-shaped animal called *Nummulites*. It is not such an impressive little object, in fact, it looks rather like a dirty, ancient coin, from which time has long removed any monarch's likeness; so it must have seemed to its first discoverer, because its name is derived from a Latin word for a coin. Those found in England are often no bigger than a penny. Humble though they might look, nummulites are an extraordinary link between southern England and Pharaonic Egypt. The pyramids themselves are constructed of limestone blocks largely made out of the skeletons of a species called *Nummulites gizehensis*. In such limestones, the nummulites may be several centimetres across, and are often weathered so that they can be seen in cross-section. Then you can see that the apparent simplicity of the fossil is misleading, because inside they are divided into numerous chambers arranged in a complex spiral pattern, which makes them in their own way as distinctive as fingerprints.

If millions of jumbled nummulites go to make one large building block the numbers that must have died to make up mountains from which the building blocks were extracted defy computation. It is even more astonishing that they are the skeletal remains of single-celled organisms. They were giant foraminiferans. It will be recalled that fossil foraminiferans are a major constituent of the Chalk – except that the Chalk species are a millimetre or less across. *Nummulites* species are found throughout France in rocks of the appropriate age and track eastwards along the Mediterranean to Egypt, and from there beyond into the Middle East and the Himalaya. The species of nummulite recovered by a collector is used to date the rocks that yielded it. Thus it is that Bracklesham (Hants) connects with the great seaway of Tethys. In the warm climate of the Eocene, foraminifera grew large and prospered, their single cells were coddled to new, brobdignagian proportions, and when they died their limy skeletons accumulated in their trillions. Children poking around in the Hampshire cliffs

257

may well unearth some Eocene nummulites, which are here close to the edge of their range. As they turn them over in their hands, perhaps wondering how money could turn to stone, they should know that they are touching the tombs of the pharaohs, and that hot desert winds etch out similar examples two thousand miles away along the valley of the Nile. The hidden landscape extends beyond our islands, because they are bounded by chance alone, and not that long ago.

The London Clay ought to be our most familiar rock, but its typical development under the capital is not exactly accessible. It comprises several hundred feet of clay, which is bluish-grey when fresh, but when weathered it is an unattractive yellowish brown. The London Basin includes the City, the West End and the suburbs and a chunk of the residential and business heart of the nation. The Chalk encloses the Basin between the Berkshire Downs and Chilterns and the North Downs, and underlies it. The bowl which holds the Tertiary in the London Basin can be drilled, not for oil, but for water, because it is an artesian basin. The Chalk holds the water, which is sealed in by the impermeable younger rocks, but which can be reached by drilling, and when tapped seeks to rise under its own pressure. This is pure water, filtered through the Chalk, and no doubt could be marketed if lightly carbonated and served with ice and a slice of lime.

The landscape of London is dominated by the more recent history of the River Thames, and particularly by the terraces which have been left by former river levels during Quaternary times, rather than by the underlying London Clay itself. Londoners have fought the Clay for generations. It is alleged to produce fertile soil, and so it seems to – in other people's gardens. Two hundred years ago, local market gardens supplied most of the city. What may have been the earliest commercial nursery was run by John Tradescant in Lambeth: where now Kentucky Fried Chicken boxes flap limply on the concrete. But in your own garden, its implacable glueyness drives you mad. It has the remarkable property of adhering to wellington boots in layer upon layer so that after half an hour of gardening you lumber about like a ham playing Frankenstein's monster. Cracks open up in it during dry spells which swallow all the water in the world; the same shrinkage causes fissures in masonry, and foundations to shift. Roses, however, love it. It seems that they like the combination of London Clay, and high levels of pollution, which discourage black spot and downy mildew. How strange that you can see the fluffy pink 'Dorothy Perkins', which is prone to mildew, in better condition in Brixton than you can weaving over a cottage porch deep in the country.

If you wish to see a passable outcrop of London Clay, you have to go to the Isle of Sheppey (Kent). This is an odd piece of England on the south of the Thames estuary, barely an island, where the Clay makes low cliffs. Mud Row is one of the hamlets, a name which captures the essence of the place. There is a makeshift quality to the houses, bungalows especially, and some of them are true innovations put up when there were no planning regulations, and as haphazard as you please. Caravans come to spend their last days on Sheppey. On the beach, the fossil beachcomber can find fruits and seeds, crabs, molluscs, and, more rarely, vertebrates. They take on a blackish, rather leaden colour from the clay. The fossils indicate marine conditions, but with ample chance for derivation of additional fossils from a land source that cannot have been far distant at the time. In Eocene rocks such as the London Clay many of the fossils still have living relatives, in a few cases even close relatives. Their ecological preferences have apparently changed little over 50 million years. The fruit of the stemless palm *Nipa* is one of the most famous of these fossils – not least because it is rather large, and not uncommon. Today, its close relatives grow in the Malay Peninsula. The same applies to several of the other fruits and seeds that are found in the clay. The crab *Zanthopsis* also has living relatives in the same area. This crab is one of the commonest fossils in Sheppey, and it is entertaining to find a chunky specimen on the beach near the insubstantial cast carapaces of its living relative, the shore crab. The rare occurrence of fossil crocodiles and turtles, and even the occasional snake, confirms the impression that the London Clay accumulated under very warm conditions. But there are a few more 'temperate' plants, also, and some of the woody fossils show growth banding typical of seasonal growth, and this has led some experts to suggest that the Clay accumulated rather further north than the tropics. Whatever the disputes about the habits of this genus or that, it is perfectly clear that the climate was very much hotter then than it is today. Long cooling since, which had already started at the end of the Eocene, has chased the flora eastwards to Malaysia, where the botanist can still collect living species. In the Eocene, South-east England sweated under humid heat, and a rich flora and fauna flourished around the shores of a relatively shallow sea in the London Basin, where turbid waters dropped their sediment to make the London Clay, and mangrove swamps fringed the seas. Land lay westwards and northwards, where the rocks thin out and there are more sands.

If you cannot see this story in the rocks exposed in London itself, you *can* go and see some spurious geology propped impertinently on top of the Tertiary. In

Crystal Palace Park (Sydenham) there is a fake outcrop built into an alpine garden. It has been built rather carefully as a series of contrasting and gently dipping 'beds'. One of the beds has been manufactured from flints, another represents a 'coal seam'. There is a beautiful little fault in the middle displacing the beds to either side quite realistically. All in all it provides a good way to demonstrate sedimentary geological principles to those with an aversion to the countryside, quite apart from the amusement of seeing coal and flints overlying what we know to be Tertiary country. This cheeky bit of pseudo-geology was designed to set the scene for dinosaurs.

Waterhouse Hawkins built the dinosaurs from a brew of cement and stone, brick, tiles and iron. They were constructed with the advice of Sir Richard Owen, the leading palaeontologist of his day, and opened to the public in 1854. The Crystal Palace – the acme of Victorian glass construction – burned to the ground, but the dinosaurs are still there. Life-sized, they loom at you through the trees from an island set in an ornamental lake. The lakeside path leads you back in time, until you reach the age of the scaly monsters. In reality you are being led back in time only to 1854; the dinosaurs are reconstructions made on the basis of the knowledge at that time. Now they look quaint and curious, but somehow still impressive, and not just because of their bulk. Richard Owen's concept for the dinosaur was based on his wide knowledge of comparative anatomy; he knew they were reptilian, and thus his vision of the monsters was, at root, a scaled-up lizard or basilisk: more massive, undoubtedly, but still distinctly lizard-like. *Megalosaurus* has thick forelegs, and rather small teeth. The Crystal Palace version was reconstructed from fragments, all that was known at the time. This version is a cumbrous animal, and very different from the two-legged, speedy, vicious carnosaurs that every child knows from the back of cornflakes packets, or the designs on pyjamas. It was not many years before complete skeletons of *Tyrannosaurus rex* were discovered, rendering all earlier interpretations obsolete.

I am glad these lumbering giants are still there. They serve to remind us of the contingency of knowledge. They provide a metaphor for progress in understanding history, an illustration of how unacknowledged concepts control what we think we know, our way of seeing, our notion of wisdom. Owen, a brilliant man, knew lizards, and he saw dinosaurs as lizards writ large. Many times in this book the hidden landscape has been revealed by a new twist in perception. Plate tectonics changed the way we see the Highlands; the electron microscope allowed us to see what really made up the Chalk; glacial theory helped us look anew at mountainous landscapes. We have not come to the end of seeing, for all that we pretend that we have seen the past as it really was. An aggressive questioner who lurks around at the end of one of my lectures on this or that aspect of the past in order to demand: 'How do you know it won't all change in twenty years?' is asking the right question.

A little further round the lake from the incorrect *Megalosaurus* is a passably accurate plesiosaur, poking its head from the water on its long and flexible neck. Back in time – further again – is a 'labyrinthodont' amphibian which is portrayed as a truly gigantic frog: this is extravagantly wrong. The model for an amphibian was a living frog, presumably, and enough was known from the fossils to be sure it was a large amphibian – but not sufficient to constrain the model further. 'Labyrinthodonts' were more like crocodiles than frogs. But in the trees you will see a bearable pterodactyl.

The London Clay covered the London and Hampshire Basins; of the two, the London Basin was the more open to the ocean, which extended eastwards into the Paris Basin. Westwards the Clay is thinner, and that is where the land lay. Through much of the Tertiary there were oscillations – a series of cycles – between the dominance of landward influence and the marine. At some times freshwater deposits of rivers, or at least more brackish water deposits, spread from the west and pushed the marine deposits eastwards, but at other times the sea invaded westwards and the rivers and lagoons retreated before it. The time of deposition of the London Clay was one of those phases when marine influence was dominant. The Reading Beds beneath are more extensively non-marine. Tertiary deposits extend westwards along the valleys of the River Thames and the River Kennet, and the further westwards the more generally sandy or gravelly they are; in the far west only little patches sitting on top of the Chalk remain – but local pits have occasionally been scraped into these outliers as providing one of the few local supplies of sand. The source of much of the Tertiary sediment must have been from the erosion of the underlying Chalk as a consequence of the uplift which followed the Cretaceous – leaving behind the hard flint, as always, to become cobbles or pebbles in gravelly rocks. But there are more exotic minerals to be found in the Tertiary sediments, such as tourmaline, which originated in the 'skarns' around granite masses much further to the west, and thus it can be inferred that the Cornish granites were being eroded, too, and contributed debris to the basin in the South-east. Other kinds of minerals seem to fingerprint a southerly origin, from the erosion of rocks similar to those exposed in the Cherbourg Peninsula today.

The oscillations that took place in the Tertiary between marine, lagoonal, freshwater sediments, and back again, make for many intriguing local differences in the succession of rock types, which have occupied the attention of devoted amateur geologists for decades. The most important contrast is that between gravelly/sandy rocks, and soft clays. The gravels and sands – if they have not been built over – produce poor soils, which support typical heaths with scots pine and birch, and with heathers and harebells, and patches of gorse. Areas of secondary woodland often include smallish oak trees and the ubiquitous sycamore, but in the areas around London the air pollution is such that the lichen decoration of the branches is less than it should be. The heathland will tend to take the high ground, with clay vales or plains between. The example every metropolitan knows is Hampstead Heath, which, miraculously, still has little bits of wildness. The view over the London Clay vale of the Thames Valley to the south occasionally stuns you with its beauty, especially at night when height and distance invest the lights of the city with delicacy and elegance, and even arterial roads can look like fairy lights. In the summer, it is possible to lie among short grass near Kenwood House; and you could fool yourself that the urban monster all around had faded as if it had never been.

The sands and gravels that underlie the walks on Hampstead Heath are highly permeable, but the London Clay underlying it is not: at the junction between the two there are natural springs where water seeping through the sands finds the surface. There are little boggy patches on Hampstead and Highgate Heath, even now, which are particularly easy to see during wet periods. Such is the evidence for a geological junction hidden away in what is almost the middle of London.

The Tertiary rocks of the London Basin form a broadly triangular area which tapers progressively from the Kent and Essex coast towards the west, and then breaks into patches beyond Newbury. Gravels and sands have been excavated commercially in many places in the Thames and Kennet Valleys to feed the endless demand for concrete and hardcore. Better still, after the diggers have done their stuff the pits can be flooded to provide a marina or a fishing-lake – or even a bird sanctuary. This is one of the rare cases where it is possible to have a commercial profit without ending up with an ecological hangover.

Woodland and heathland, associated with underlying sands and gravels, occur over stretches north and south of the Thames: Hayes Common, near Bromley; the common at Beaconsfield are examples. Small Tertiary gravel patches have survived on top of the Chalk in the Chilterns, near High Wycombe, and near Hemel Hempstead, which prove the erosion that took place between Cretaceous and Tertiary. Areas underlain by the Bagshot Sands are singularly infertile. These and similar beds occupy extensive areas around Ascot and Camberley. Because the soils on top of the gravels are, however poor, lime-free, this is the country where the rhododendron (*R. ponticum*), recognised by its rather insipid lilac flowers, has become naturalised to excess; indeed, it can almost be used to make a geological map of the Tertiary sands and gravels. Apart from when it briefly erupts into flower in June it is a sorry plant. Nothing grows beneath it – not even fungi. It out-competes everything. Its evergreen, leathery, dark green leaves induce a depression of the spirits; even birds avoid it. In the archetypal Tertiary gravel garden other *Rhododendron* and *Azalea* species hold sway, the former again brief-ly glorious during flower, the latter with the added virtue of deciduous habit. I have to admit that the kind of drift of species seen in the Royal Horticultural Society garden at Wisley cannot fail to inspire admiration, but a certain scale is necessary to fully appreciate rhododendrons which the suburban plot sometimes lacks. Miniature azaleas seem to be more the right size in this context. The other regular feature of such gardens is the heather and conifer bed, whereby a peat bed is built up to combine dwarf cypress having ornamental foliage or habit – glaucous or prostrate, variegated or fastigiate, according to taste – with species of Ericaceae, to fashion an idealised, miniature heathland. The fortune of many a garden centre has been built upon the heath bed. It is a curious garden convention in the soft South-east, because it seeks to emulate floras that only partly belong there – many of the azaleas would be happier with a moister, Atlantic climate. The ecological tastes of these plants do reflect the geological landscape – so much so that one can guess the geology from a passing car if the appropriate garden design is present – but the taste of the gardener is as important, and assuredly more mysterious.

The gravels alternate with clays in some areas, and there has been quite exten-sive local brickmaking, producing red bricks around Reading, for example. Other clays were particularly suited to making long stemmed churchwarden's pipes that were smoked by many, many more professions than the churchwarding one in the last century – it is impossible to dig any garden without finding the broken stems of these pipes. They were manufactured (and broken) in truly nummulitic numbers. Their use fell out of fashion when it was realised that there was a close connection between use of churchwardens and lip cancer. The fragments are not uncommonly mistaken for fossils, and it is true that they have a passing resemblance to the stems of crinoids (sea lilies). Around Wareham (Dorset) the pipeclays in the Bracklesham Group are still exploited commercially in the ceramics industry.

Although most of the Tertiary rocks are soft enough to scrape out with bare hands, there are conspicuous exceptions. Some rocks dating from the lower Tertiary have become *indurated*. This means that a cement has bound the particles of the sediment together; the toughening this causes is remarkable. The cement is silica, which is the same material as flint. Indurated rocks are just about as hard and durable, too. In the old village of St Michaels abutting St Alban's (Herts) there is a large block standing near the ancient ford which looks at first glance like a lump of concrete carelessly dropped by a cowboy builder. But it is an entirely natural lump of conglomerate, a good specimen of Hertfordshire 'Pudding Stone'. The 'plums' are, of course, plumply rounded flints, often attractively colour banded in pinks and browns, while what cements them together is a white to yellow quartzite. It is an almost indestructible rock (far harder than most commercial concretes) and so was picked to give strength to cottage walls, or pieces found employment as mounting blocks for horse and carriage – the St Alban's one would do well. Even more indestructible are the indurated quartzites known as sarsen stones. They were mentioned as comprising the 'uprights' at Stonehenge, and as making up the stone circle at Avebury (Wiltshire), both on the Chalklands. But they are remnants of Tertiary beds that once more extensively overlay the Chalk. Their toughness has meant that they have survived, while all the softer rocks around them have been eroded away. They are composed of grey, pinkish or yellowish quartzite, and blocks come in sizes ranging from fist-sized to several metres long. The fossil remains of tree roots can be seen in some of them, which usually show as tubes a centimetre or so in diameter. From these it can be concluded that the sandstones which became hard quartzites once they were cemented were originally of non-marine origin, like most of the western Tertiaries. Usually, sarsens are scattered here and there over the surface of the Chalk, especially in Berkshire and Wiltshire. Occasionally, they occur in numbers on the Chalk Downs in a 'valley of the rocks'. One valley at Ashdowne House (Oxon) I find remarkably eerie, as if Dame Barbara Hepworth had abandoned her half finished works to a design known to the gods alone. Blocks lie here and there as they might if they were waiting to join a stone circle.

Hertfordshire 'Puddingstone', a tough Tertiary conglomerate made of flint cobbles bonded by quartzite and which is not unlike concrete in appearance. This block lies in St Michael's village at the edge of St Alban's, Herts.

The Tertiary rocks mentioned so far are Palaeocene and Eocene in age. The formations follow one another so that they are easy to read from the geological map: above the Chalk, The Reading Beds, then London Clay, Bagshot Beds, Bracklesham Group, Barton Formation, Bembridge Beds, and so on. But there are many local variations which are not pursued here. Nonetheless, it is possible to spend an enjoyable day or two with the British Geological Survey Regional Geology guides following the succession of beds in Bracklesham Bay or around the coast of the Isle of Wight. At Whitecliff Bay at the eastern end of the Isle of Wight the succession of rocks above the Chalk can be followed north of Culver Down, a repetition of the story at Alum Bay. You move rapidly up through the sequence northwards, and as you do so the marine influence on the sediments is eventually replaced by non-marine: fossils of leaves and roots and insects mixed with clams and snails have suggested lagoons or rivers. The steep dip of the rocks soon relaxes towards Bembridge. The rocks fringing the sea around the town are quite well exposed on the foreshore. The Bembridge Marls are one of the few deposits in which insect fossils are rather common. Considering their absolute dominance of the living world in numbers of species, we know very little of

Stonehenge, Wiltshire, the 'uprights' are sarsen stones of Tertiary age, which have been used in stone circles and burial chambers very widely in southern Britain.

their history. There are more species of beetles than anything else. Why are beetles so successful, and when did their dominance start? Nobody knows, and there are hardly any scientists studying the question. Perhaps there is no money to be made from the answer.

It is easy to spot fresh- or brackish-water snail fossils in the limestone ledges near the Foreland. The rocks here are nearly horizontal, so you walk over ancient bedding surfaces on the foreshore. Keep on walking around the coast to Nettlestone Point, and you can enjoy a rest from geology over a good lunch in a slightly raffish seaside pub that does marvellous *moules marinières*. The youngest Tertiary beds around Parkhurst Prison and Hamsted are Oligocene in age, with more molluscs and plant remains. But a lucky collector may find the bones of a mammal or a crocodile.

I must mention the curious little outlier of Tertiaries – marooned among older rocks – which occupies the flat-bottomed valley around the town of Bovey Tracey in Devon. These rocks accumulated in a large lake in the latest Eocene to Oligocene, and are remarkable for having beds of lignite composed of the gigantic conifer *Sequoia*, which is today one of the tourist attractions of California. It must then have flourished in the West Country.

There was a period of widespread uplift and erosion across Britain during the Miocene. Because this was also a time of active mountain building in the Alps the hiatus in the British Isles has been regarded as a kind of distant shrug attendant on a major convulsion. The next rocks to be deposited – the Crags – lie uncomformably on an eroded surface of Chalk or earlier Tertiary. The fact is that the Miocene has been regarded as missing. Britain's wonderful geological record has only this one major hole. It is a substantial gap as well, because there are thick Miocene formations elsewhere in Europe, and it was an important time in the history of life. It marks the beginning of our modern mammal fauna; earlier mammals included many archaic beasts with few close living relatives, but from the Miocene onwards most of the mammals would look familiar enough for the average naturalist to have a guess at their living relatives. It does seem possible that there might be the merest *taste* of the Miocene in our islands, because there are some curious rocks in East Anglia known as the Boxstones, including moulds of molluscan fossils which, my molluscan friends tell me, are now most likely to be of Miocene age, although they have been claimed as anything from Oligocene to Pliocene in the past. I hope my friends are right, so that our British sedimentary geology might be truly compendious. The Boxstones are brown sandstone cobbles found in the nodule bed at the very base of the Crags in Suffolk. This bed marks the resumption of sedimentation after the Miocene 'gap', and the Boxstones are the remains of an unknown deposit that was once more extensive, reduced now to a handful of relics.

At this point it is necessary to take a jump, far northwards, back to western Scotland, because this is where the other Tertiary rocks in Britain are to be found, although they could scarcely be more different from those of South-east England. Extensive formations of clay are not to be found there, nor are there sands and gravels laid down in vanished rivers and lagoons. Nor is there a soft landscape that might go with such rocks. Instead, there are lavas, black and uncompromising lavas, fretted outpourings from the belly of the Earth.

Here in Scotland is a repetition of history, another cycle initiated. It will be recalled that in the same area there was another great ocean four hundred and fifty million years before – Iapetus – with nearly the same course as the Atlantic today. The Inner Hebrides record, in our own islands, the widening of the *pres-*

ent Atlantic Ocean, the beginning of the paragraph in the great narrative of the Earth at which we now find ourselves. This is the phase when Europe and North America drifted apart, driven by the great Earth engine in the depths. Volcanic rocks are the legacy of this phase. Ocean widening is written upon the Earth in volcanic rocks, because it is lava which solidifies to dark basaltic rock to take up the 'gap' between separating continents. Fire forges the shape of the world, because this is how oceans are built, from strips of volcanic rock. The process continues. In the middle of the Atlantic ocean new volcanic crust is *still* being added as the Eurasian and American continents separate. If you wish to find landscapes reminiscent of Northern Skye it is to Iceland you must travel, which sits astride the ridge at the centre of the Atlantic Ocean where crust is generated today. Here the lavas have barely cooled, indeed still flow from time to time, and steaming springs let you test the reality of deep heat with your hands. North-west Scotland and Antrim have long been left behind as the centre of such fiery construction, because the hot centres of lava production moved to the west. But the signature of its history is graven upon every volcanic rock. In the Inner Hebrides some of the oldest rocks in Britain and some of the newest are juxtaposed, the metamorphic Moines or the Dalradian from before the Cambrian cheek by jowl with the upstart lavas. Both are linked by the deep alchemy of crustal processes, like two strangers of different generations – thrown together by apparent chance – who might yet acknowledge a mysterious affinity.

Tertiary volcanic rocks extend from Skye in the north, southwards through the islands of Rhum and Eigg and Muck to the peninsula of Ardnamurchan; thence southwards again to the great Isle of Mull; dykes cross Islay and Jura and Kintyre and all these dykes point further southwards again, to Arran, which has already been visited in this book. Thence to Ireland, where the Tertiary igneous rocks underpin Antrim, and Slieve Gullion and the Mountains of Mourne. What we see now as nothing more than a contribution to our landscape once darkened the sky with ejecta and smoke, while rumbling eruptions stirred up ancient memories in deep faults inherited from Caledonia.

The Giant's Causeway lies on the north coast of Northern Ireland. A giant rather than the devil is associated with this famous piece of geology, which is perhaps surprising considering the fiery origin of the rocks. Great floods of dark basalt extend over 3000 square kilometres in Antrim, flow upon flow. Some individual flows extend for up to ten kilometres, and can be more than thirty metres thick; most are thinner. Basalt lava welled up through cracks in the crust, or through volcanic vents. The Causeway shows what happened as the basalt cooled: it contracted into polygons – like a honeycomb cast in stone. Many of the polygons are hexagons, one of the commonest figures in nature. The hexagons are arranged in columns, crossed transversely by contraction cracks. Weathering along the cracks generated the steps of the Causeway. Elsewhere, there are organ pipes, or harps, or chimneys made out of columns in different combinations or sculpted in several ways by wind and weather. West of Mull the Isle of Staffa, which inspired Mendelssohn to one of his filigree confections, is another celebration of the possibilities of ranks of columns: scarcely a hidden landscape, more a blatant display of the physics of thermal contraction.

All these extravagancies originated as lava flows, runny lavas somewhat akin to those still spewing out regularly in Hawaii, with low viscosity so that they flowed for long distances. If it does not seem a contradiction in terms, these basaltic eruptions were comparatively peaceful, that is, the lava flowed without being accompanied by massive and catastrophic explosions. Most of these basalts are finely crystalline because they cooled too rapidly for larger crystals to grow; but some are porphyries, which have a few large felspar crystals (which presumably grew at depth) floating in basalt like cherries in a cake. Between flows there were quiet periods, long enough for soils to have been produced by the weathering of the fresh lava. These kinds of lavas decomposed to give a rather rich soil, sufficiently so for trees to have flourished. The next flow would have fried them. But their former presence is proved by the charcoal smears that remain, set beneath the uncompromising gravestone of the overlying irresistible flow. There is a famous example of a tree, as high as a house, buried in basalt – fossilised by fire, as it were – south of Rudha na h-Uamha, west Mull. The volcanic piles are thick. Ben More on Mull tops a pile of 1000 metres of lavas. The bareness of the landscape suggests an antiquity as great as that of the Outer Hebrides, which shows how geological complexion can deceive. Flow upon flow are seen wonderfully well in northern Skye, and on the small island of Canna (Inverness), and on Eigg. The landscape is stepped; inland the several flows show as horizontal ribbing in the hillsides – but the crowns can be as flat as the upper-most flow. Basalts of this kind are described as plateau lavas because table-top mountains are the characteristic landscape. The 'crusts' of the flows are the more easily weathered, compared with the fresher, denser bases, and it is this property that creates the steps in the landscape. On Skye, the weathering of basalt in The Storr has produced a fantastic array of towers and spikes a hundred feet high, or more, dubbed with names like the Needle, or the Sanctuary, all presided over by the Old Man himself, a towering monolith fifty metres high. The Sgurr of Eigg, that island's crowning crag, is a sheet of pitchstone, a volcanic glass frozen from lava by still more rapid cooling. It looks like black glass, too, and breaks into shards as sharp as flint.

It is astonishing to realise that the plateau basalts we see today are only remnants of what was erupted during Eocene to Miocene times. The vast Atlantic deserves such a memorial on our shores.

The centre of Mull around Beinn Fhada to Loch Ba is the site of a huge caldera – a vast circular area of subsidence where the earth sank down after the trauma of eruption. Water filled the caldera to form a lake, and subsequent eruptions even spilled out lava under water, where it solidified into 'pillows'. New volcanoes built up within the crater; great volcanic cones must have towered over the Scottish landscape, of which we see now only the exhumed remnants, deeply eroded to their roots. All the geology around this great extinct volcano seems to run in circles, or rings around the volcanic centre; especially dykes, which run in profuse, concentric sheets from the centre, the frozen legacy of lava intruded into the country as the ground bellied up, engorged by the great plume of heat from below. There are varied rocks in this volcanic region (they are often called volcanic 'complexes' with good reason) – far more than the variations on the theme of basalt displayed by the black plateau rocks. On Mull, there are rocks which record eruptions from no less than three successive calderas.

The magma chamber from which the eruptions derived was like a great cooking pot of liquid rock from which different siliceous brews could be squeezed forth. Early crystals settled to the bottom of the chamber in the same way that lentils and pearl barley settle in a stew pot. With early brews removed, what remained was enriched in some ingredients, impoverished in others, so that new eruptions produced yet further species of rocks. At depth, some coarsely crystalline rocks solidified slowly: dark gabbros, rich in the mineral olivine. Later volcanic rocks cut earlier rocks, dykes cut through joints in solidified eruptions . . . And then some gummy lavas were brewed that blocked up volcanic vents, and were likely to have been blasted out with great explosive force – like champagne uncorked. These are now immortalised as plugs and breccias (often called volcanic *agglomerates*) on the flanks of Ben Buie: rough and rubbly, they look like hardcore dumped by a giant. The thicker brews are lighter coloured rocks – even creamy – because the minerals change with the cooking, and have little in common with dark basalt (on the geological map these pale rocks appear under names like felsite or rhyolite). The Mull Tertiary volcanoes are a mass – one might even say a mess – of different rocks, interleaved with the complexity of a piece of eighteenth-century marquetry inlaid upon a circular table. This is the last, and best-preserved visit of Vulcan to our islands. The complexity of the geology, in conjunction with the moist climate, may account for the fact that Mull has the richest fungus and lichen flora in the British Isles.

At Ardtun, on the Ross of Mull, there are some sediments interspersed with plateau lavas, and quite exquisite remains of leaves are preserved in them. They look as if they were drawn in sepia – complete with veins – upon the pallid, shaly rock by an artist with an extraordinary gift for verisimilitude. The leaves were obviously derived from broadleaved trees, and so would tell the finder instantly that these lavas were erupted when the flora had a 'modern' look, and likely after the Cretaceous. The commonest leaves are probably those of plane trees (*Platanus*). But there are also maidenhair trees and hazels and oaks which suggest a climate warmer than it is today.

As on Mull, so it is in Ardnamurchan: another great, extinct volcano and most of the rocks arranged in rings around it. Just about the whole tip of the peninsula – more than five miles long – can be understood as the dissection of a vanished fiery waste, with several successive volcanic centres. The concentric rings which construct the landscape are, if anything, clearer here than on Mull. Dark igneous rocks support sparse vegetation; rounded bluffs often carry barely more than a few encrusting lichens except where joints have eroded into narrow gulleys. Harder rocks take the higher ground, while streams eat out, or bogs settle upon the softer products of the magmatic cooking pot. Dykes stand out as low ribs. There is a ring of 'eucrite' hills three miles across on the north-west side of the Peninsula that writes out the geology upon the ground as clearly as any geological map: Meall an Fhir-eoin, Meall Meadhoin, Meall an Tarmachain, Beinn na h-Imeilte, Meall Sanna, Meall Clach an Daraich make a ring-o'roses about the centre of the ancient volcano, and from the top of one of these hills yet further circlets related to different volcanic rocks can be guessed or seen. Imagine, too, that the whole volcanic pile originally lay far above your head; time and erosion have scoured it down to where you stand today. But elsewhere, the subsidence has preserved rocks which might otherwise have been eroded away. If you wish to see volcanic agglomerate go to Maclean's Nose on the south coast seaward of Ben Hiant. This is a great mass of lumpy filling from a hole blown by a massive explosion – the filling is tougher than the rocks that surrounded it and now forms a striking cliff: time has reversed the original relief. Who Maclean was is not recorded, but his nose must have been colossal.

The greatest peaks among this scatter of Tertiary igneous wonders are, as always, huge igneous intrusions, and the greatest of all these are the Cuillins of Skye. The Black Cuillins are made of gabbro, black or darkly green, with crystals as coarse as granite, but dominated rather by dark minerals. Gabbro is not a common rock in Britain, but there is plenty in Skye, soaring above the plateau basalts that make up the north of the Island. The same rock types are found on Rhum. No doubt the dark colour contributes to the peculiarly brooding quality of the Black Cuillins, but the mountains are also considerable enough to have been much sculpted by ice, which has scoured out great corries separated by ridges of uncompromising jaggedness; these would test the nerves of a stronger climber than I. Snow lingers on Sgurr Alasdair, the highest peak on Skye; Inaccessible Pinnacle nearby is one of those names that requires no descriptive gloss. Can there be any more grimly impressive place in the British Isles than Loch Coruisk? The glacier that filled loch and corrie ten thousand years ago seems to have left only yesterday so fresh the rock remains; the corrie often steams and smokes with mist, and then the onlooker might be fooled that *here* was the volcano, rather than hidden among the gentler slopes of Ardnamurchan. How ironic that the 'crater' is the work of ice not fire, and that the sculpting was carried out no time ago, geologically speaking, for all that Loch Coruisk broods an atavistic nervousness that speaks of deeper time.

The Red Hills to the north and east of the Cuillins provide a striking contrast: for these are granite-like rocks – and pink felspars provide the basis for their name, just as for the colour of the hills when the sun is low. They are late products from the cooking of magma, near the end of the refining process. Around Beinn na Dubhaich, as they were intruded granitic rocks baked and altered Cambrian to Ordovician limestones of the same kind as those I described in some detail from around Durness. This is one of the very few places in the British Isles where heat has produced true marbles, the metamorphic kind. Goat Fell, on Arran, is another of these Tertiary granites; an almost perfectly circular intrusion is the high point of the island, a brisk walk upwards from all sides. Wherever such tough granites reach the surface they resist the inroads of weathering. The cliché about faces set like granite against adversity is not inappropriate.

The islands of St Kilda lie fifty miles west of the Outer Hebrides – so remote

Crofter's cottage, Skye. Local materials were recruited to make these simple dwellings, which remain part of the landscape.

that they have a subspecies of wren all their own – and these desolate rocks are another defiance on the part of Tertiary fire-formed concoctions against the worst of the elements. The cliffs plunge so steeply into a sea that is perpetually on the boil that it is a wonder that the place was ever inhabited, but so it was, and over centuries. Finally, and almost ludicrously alone, there is Rockall, 190 miles west again of St Kilda. A mere seventy feet of granite breaks the surface, buffeted continuously by mid-Atlantic swell; the last redoubt, the edge of the nation. Though so remote, it has been known to have been composed of granite since 1814. Few people ever see it, and perhaps rather few want to. One man has recently lived on it, for a while, an achievement which is almost endearingly bonkers.

Dykes, like bees, come in swarms. Along the foreshore on Arran dykes run out into the sea with the regularity of groynes; they often form upstanding walls because they are generally tougher than the surrounding rocks. They make the natural 'roadways' for children to run along to reach the waves. There are more than five hundred such dykes on Arran. They are vertical, as are virtually all the Tertiary ones: they slice clean through the geology of the country rock. On a close look, the careful observer will see how the margins of the dyke are 'chilled' by contact with the cool rock: this in turn is often discoloured where it has been baked by the insertion of hot magma. A dyke may be a few centimetres wide, or as wide as a room. Dykes have been intruded at any of the times when volcanoes have scarred our landscape. The ancient dykes of Scourie are among the oldest rocks in the country, but the Tertiary dykes of Scotland are among the youngest. The latter swarm well beyond the Scottish volcanic centres: some even pass into England. They run straight for miles. The trend can be seen on any geological map: North-west–South-east, almost at right angles to the celebrated Caledonian direction. The igneous rock that forms the dykes is usually a dark, dense stone in which it is difficult to pick out crystals. These are best seen where the sea has polished the surface: nowadays, there are a number of different names applied to the rocks that make the Tertiary swarms ('quartz dolerite' is one of the commoner ones). Some dykes were no doubt 'feeders' for lava fields, now long since eroded away, but most filled the 'cracks' that bellied outwards as hot magma dilated the crust beneath. There must have been a lot of distension, too, if one adds together all the bees in the swarm.

On Skye there are a number of crofts maintained in good condition, and many more rather splendid in decay. This is true of Trotternish in the north, and of Kilmuir. These single storey, long, low houses are built from the local stone – after all, there is always plenty to choose from – as are the dry stone walls that circumscribe the plot, and any outbuildings; there is usually little attempt to dress the stone. A chimney at either end, a thatched or tarred roof, a root cellar perhaps, a few trees planted for protection, and that is about all there is to it. But they do seem to belong well in the landscape, and in igneous country one might say that they, too, were brewed from magma in the refiner's fire deep in the Earth.

The Tertiary set North apart from South almost as in the time of Iapetus. Lava spewed from vents, or volcanoes blasted the sky in the North-west, as a warm sea fringed with rich forest thronged with fish and crabs in the South-east, while there were rivers, lakes and lagoons, further to the West.

17

East Anglia: Sky and Ice

EAST ANGLIA is, I suppose, something of an acquired taste. It is both the flattest and the newest part of England. The geology hides beneath low heathland, and endless arable fields. Inland, much of East Anglia is the nearest thing we have in the British Isles to the grain belt prairies of the United States. Field boundaries and their hedges have been sacrificed to the interests of mega-farming. The remaining verges have been sprayed or fertilised, and either course of action leads to 'green deserts', where a few tough grasses like *Dactylis*, and a few tough herbs, like nettles and hogweed and cow parsley, do an inadequate paint job over a deprived coach-work.

This description is, thank goodness, an injustice. It could, and does, apply to hundreds of square miles of East Anglia. But there are corners of north Essex, Cambridgeshire, Suffolk, and Norfolk which have a charm of their own, and which have retained a character that derives from the hidden landscape as much as any other part of the country. There is not much in the way of hills, it is true, but there are villages in which Elizabethan buildings are the norm, and not fussed over like pampered poodles; there is the most infuriating maze of minor roads anywhere, with signposts that promise destinations which never materialise; and there is a geological narrative that takes us through the last few million years more completely than anywhere else.

A broad bank of the Chalk passes through the middle of East Anglia, but the most relief it can manage is the Gog Magog Hills, which are charming and typical Chalklands, although scarcely precipitous; further north, in the Breck country, its impact is muffled by sands smeared on top of it. On the coast, one can see erosion writ large, but one can also see sediments being laid down, and appreciate how cockles and green shore crabs will become the fossils of the next millennia. This is a fragile part of the country: nowhere here can be found indestructible gneiss or enduring rocks tried through time and indurated through depth of burial. It is perfectly possible that the whole rock record will simply be eroded away. It is a place to feel the ephemeral grasp we have on the past, how its endurance is a matter of chance, and how the history we do know could have been erased if only erosion had proceeded just that little bit faster.

The narrative of the Tertiary in the South-east paused at the Miocene 'hiatus'. It begins again in the Crags of East Anglia, which overstep the earlier deposits, and mark Pliocene sedimentation in what might be considered as the precursor of the present North Sea. 'Crag' is an East Anglian name for a shelly gravel, which will do very well for a brief description. The Coralline Crag is the oldest formation, and occupies the subsurface over a small area of Suffolk between Boynton and Aldeburgh; in fact, outcrops are very hard to find, but there are a few remaining along the Butley River. The formation is a stupendous misnomer, because corals are comparatively inconspicuous fossils in these shelly rocks. The 'coralline' refers to an abundance of sea mats (bryozoans), some of which, like *Meandropora*, can have a superficially 'corally' look about them. Sea mats are commonly washed up on the shores of Suffolk today – they often have the appearance of a kind of crispy, white seaweed, and I suppose are dismissed as such by many without a second glance. Parts of the Coralline Crag are just made of bryozoan fragments, and the oblique bedding shows that the formation was deposited under the influence of strong currents: it is really a series of submarine bryozoan dunes. If you are lucky enough to find a new exposure, a range of interesting fossils can just be plucked out of the unconsolidated rock – perfect, apart from yellowish staining. Clams, snails, barnacles, and, most interesting of all, a giant brachiopod, *Terebratula maxima*, the last relic in the British rocks of a group of fossils that roofed Snowdon (Ordovician), formed hummocks in the Carboniferous, and were as common as any animal in the Chalk. The liminess of the Coralline Crag makes for an interesting flora, too, and I have found wild mignonette, musk thistle, and my first and only discovery of the cut-leaved self-heal on this little patch of Crag: chalk flowers all.

The rich faunas of the Coralline Crag are usually taken to indicate a warm climate – warmer than it is today – and I like to think of it as a kind of Mediterranean paradise, which slipped, gradually, towards the glacial age.

The Red Crag follows the Coralline Crag, but has a much wider area of outcrop (often directly overlying the eroded surface of the London Clay), and hence records a shoreward invasion of the sea. Sands are typical, which were deposited by marine currents that have left the bedding tipped and curved. The red colour of parts of it is, as usual, the result of staining by compounds of iron. Exposures are not hard to find. Seashells of many kinds are common at certain levels, and particularly snails that look for all the world like whelks. This is scarcely surprising, because they *are* whelks. For the first time in this book, some of the fossil species plucked from the cliffs are exactly the same species as those now living. The familiar is starting to oust the unfamiliar. There are winkles and cap shells, too, and clams that you might pick up on a sandy beach, and mussels and oysters. Were it not for the staining you could easily mistake these fossils for shells dead for a few months. Successive levels in the Red Crag record a cooling climate: the species that preferred life in the warm sea water of the Coralline Crag disappeared, to be replaced successively by species that preferred cool, and then cold seas. In this modest fashion was the great Ice Age announced.

East Anglia preserves the fullest narrative of the Pleistocene in Britain – the most recent, yet one of the most dramatic episodes in our long geological history. In many other parts of the country the glaciation was simply too overwhelming: the only rocks preserved are those left behind *after* the glaciers retreated, but the successive pulses of the occupation of the ice have left little record. In East Anglia, sediments continued to be laid down in freshwater sites and in estuaries; patchily, of course, and with breaks here and there, but in Suffolk and Norfolk the transition from prehistory to history is encoded in the rocks. These young rocks run along the coast from North of Aldeburgh in Suffolk to beyond Cromer in Norfolk.

Man had already appeared, spread, invented a variety of tools, learned to draw, hunt and perform any number of tricks by the time the climate cooled in Norfolk. Our species had emerged in Africa about 1.5 million years ago (the most accurate date as I write), but that is not part of our story. The only find purporting to be the crucial transition from ape to man within our islands was the famous forgery from Piltdown. Actual bones of early humans are very rare in Britain; their knapped flint tools are commoner. The species no doubt reached Britain from the continent by way of a land bridge that connected us to France, at the end of a long, long trek from Africa. These early tribes would have followed game, or elephants, or bison – French and Spanish cave drawings of marvellous vigour and economy show how these animals were revered, beyond their capacity to provide calories and skins. Physical anthropologists are an argumentative lot, even by the standards of most scientists, and I doubt whether there is any aspect of early human history that is free from controversy. Neanderthal Man, who was associated with glacial sites, is regarded by some anthropologists as a race, by others as a distinct species who was displaced, or perhaps even hunted down, by us modern humans. The raw material for such controversies are bones, and bones of humans are rare.

Every palaeontologist would like to find a fossil man. Usually, it is the tough skull that survives, or teeth, or a piece of the jaw. This alone can be enough to ensure a clear picture of the kind of human from which it came. At this point I should admit to my own triumph of optimism over judgement. Several years ago, in the Suffolk seaside town of Southwold, I overheard a young woman in a shop say how her husband had found a skull in the cliffs nearby. The cliffs near Southwold are just the right age to have very early human remains, and other mammals are known as fossils there. I imagined the front page of *Nature* reporting the discovery. Before long, I was in possession of the precious fossil – a cranium, obviously human, and as ancient-looking as you could wish. Was it just imagination, or did it have a slightly protruding brow ridge? I carried the remains to my colleague Chris Stringer in the Natural History Museum. What was slightly wounding was the speed with which he pronounced it 'modern' . . . he might, one felt, have prefaced it with something like: 'My goodness! what an interesting find!' before passing judgement. Apparently, human skulls are not as uncommon as you might suppose on this eroding coast – ours might have been the mortal remains of a medieval sailor, or a fragment from a vanished graveyard. However the discoverers did actually find a perfectly genuine fossil elephant tooth a year or two later.

The charm of geological discovery is partly serendipity. In other sciences, experimenters set up complex arrangements of particle accelerators and batteries of detectors to catch a fleeting entity that a dozen people in the world might really understand. Millions of dollars are spent in the search. Theoreticians propose models; thinkers brew up new abstractions; experimentalists devise critical tests. But the history of humankind can be overturned by a schoolboy idly picking at a cliff when he should be studying for his mock A-levels. In the odd system of values in which these things are reckoned this implies that historical science is regarded as a lesser science – 'stamp collecting'. It is rather as if all literature except

Shakespeare and Proust were dismissed as inadequately grasping the complexity of what it means to be alive. Perversely, geology suffers from being too readily explicable, and by being too wedded to chance discovery.

The record of the Pleistocene Ice Age is preserved thanks to the vagaries of chance. After all, ice erodes rather than constructs, so what you see of its former activities is either the damage it has inflicted, or the sediment which accumulated at the margins of any ice sheet. What shall we know of the appearance of Greenland once its huge ice sheet has melted? Animals, humans included, crawled into caves to escape a little of the fierce Pleistocene cold; they brought their prey there to butcher it or gnaw on the bones. Where chance allowed the drip, drip of calcium-rich waters to fall upon waste fragments – bone shards, teeth, broken implements – they stood a chance of being preserved beneath stalagmite, and then another chance of being discovered many thousands of years later. Hyenas, and bears and wolverines gnawed and chewed in different ways. Possibly once one of these gnawers was frightened away by the approach of some other animal and left a more complete skeleton than it usually would have done. So chance dances with chance to create a pattern which an investigator can eventually discover. Traces of fossil human blood on a stone hand axe have just been reported, thus opening a whole new field of possibilities in palaeo-pathology. Maybe more chance discoveries will soon permit us to see bits of genes of Neanderthals? It will depend on the intelligent interpretation of what chance leaves behind.

The Ice Age is not a single age at all, but a series of cold periods separated by times when the climate was as warm, or even much warmer than it is today. Glacial periods alternated with interglacial: but 'Hot and Cold Ages' does not have the same drama about it as 'Ice Age'. It is known that even upon the greater cycles there were smaller cold-warm climatic fluctuations, described as stadials and interstadials. The historian focuses his microscope on ever finer detail, as more precise methods of analysis are invented; he discovers more, and racks the focus up yet again, which suggests further questions, and so on.

It is not so long ago that the influence of ice upon our scenery was denied by many reputable observers: other explanations were in currency for the scratches on Scottish boulders, or what we would now recognise as fossil moraines, or scraped and deepened valleys. The clues that allowed recent description of the sequence of events through the Pleistocene were derived from deep-sea cores of sediment. These were obtained many miles away from the direct influence of ice, from oozes underlying quiet ocean wastes. This is where chance could be cheated. A steady and unbroken rain of shells of tiny foraminiferans falling to the sea floor faithfully recorded climatic change. In the cold phases, species of foraminifera were different from those in the warm phases; there are also changes to the isotopes in the crystals making up the shells. I grew up with five cycles of glaciation named in the textbooks, all of which were named after tributaries of the Danube — Donau, Gunz, Riss (=Saale), Mindel, Wurm in ascending time order — but thanks to these new oceanic data the number of cycles is now known to have been many more. Every year, it seems that more wiggles are added to the temperature curve for the last two million years as the investigator focuses more and more clearly upon the past.

Back in Norfolk, clever and assiduous people had for decades pieced together a Pleistocene narrative from coastal cliff sections, from pits inland, from temporary road cuts — wherever chance allowed an exposure. Elsewhere, cave floors were exhumed, such as Kent's Cavern in Devon, or the famous chambers at Cheddar and Wookey, or Minchin Hole in Gower.

As the ice sheets waxed and waned, so the sea fell and rose. Great rivers like the Thames responded; at glacial times they were rejuvenated to cut downwards, but when the sea level rose again as the ice melted they became leisurely and swampy, and deposited spreads of silt and sand. This history is preserved in a succession of terraces, eight or more, running alongside the Thames. Even the most gentle hill in the middle of London indicates one of these terraces. The most famous terrace of them all has yielded fossils of mammoths and hippopotamus and tiger that flourished in the great interglacial when the climate was at its warmest. One of my favourite old exhibits in the Natural History Museum showed the bones recovered from near Trafalgar Square, and I liked to imagine how the ghosts of these great and vanished animals might emerge at night and stalk around Nelson's Column, or pass up the Charing Cross Road ignoring the theatrical reviews. You do not need to find hippopotami to tell you that the climate was warm, nor woolly mammoths or wolves to tell you when cold periods were dominant; fossil pollen can tell almost as much about the climate; so can beetle wings, and other such minutiae. But mammals dramatise it.

As the ice sheets waxed and waned, so the very land fell and rose, because the weight of thick ice itself depressed the crust. Beaches cut when the land was depressed were raised when the land rebounded after the ice melted (like a mattress relieved of a heavy body). Raised beaches preserve other chance records of the Pleistocene, particularly in Scotland and Wales where the ice was thickest. All around the Scottish Isles terraces were left by the sea at a hundred feet and twenty-five feet above where the waves lap on a calm day today. Caves which were once at sea level are now beached and dry. Limestone caves in the Gower Peninsula are important examples for the history of Pleistocene environmental change.

So Ice Age records are profuse, but diffuse: caves, rivers, seas, swamps, lakes, peats, all doing different things at the same time in different parts of the country. Time correlation between one area and another is difficult, and remains so in spite of recent technical advances. Does a 'warm' collection represent one interglacial or the next? Or maybe an interstadial has been underestimated and was warmer than we thought? It is clear that the final ice advance was not the greatest, for older glacial deposits extend further south than younger ones: it is as well they do, for otherwise some part of our history would have been completely obliterated.

If the bedrock is the entrée of the hidden landscape, the Ice Age provides the seasoning. Hardly any corner of the country was untouched by its influence, and it has already been mentioned many times in the course of this book. In heavily glaciated areas ice and frost sought out the weaknesses in faults, or scoured softer sediments. The deep gouging of the Caledonides was the latest episode in their long exhumation. At its maximum, ice cover left untouched only the southern-most part of England, and even there the frigid conditions – like the tundra today – affected every aspect of erosion. Ice diverted rivers, and dammed lakes. The Vale of Pickering is as flat as the fens because it is the floor of an Ice Age lake ponded by a glacier ice which spread outwards from the North Yorks Moors.

Visualise a great ice sheet moving from the North: it retreated and advanced again repeatedly, while all the animals and plants followed the climatic shifts back and forth across Europe to keep abreast of their natural habitat as it moved around; our nascent islands were still attached by an umbilical cord of land, so that Europe's history was closely tied to ours. In glacial phases, glaciers deepened valleys which would become lochs, scratched boulders, and scoured mountains which would now be scalloped with corries or cwms; prolific sands and gravels washed out from the melting glacier front at every spring. Peat accumulated in bogs, and preserved the pollen of a rich flora.

Then, once again, the ice sheet advanced and obliterated much of what had been left behind before. It carried blocks of rock far from their origin in outcrop, which after the ice retreated once more were dumped as exotic erratics in some surprising resting-places*. On the Carboniferous Limestone in Yorkshire there are propped erratics of Shap Granite, or Silurian grit. Eventually, the great glaciers retreated, dumping material in moraines of a dozen different kinds, which are composed of rocks which still look all jumbled up today, having cobbles of all sizes and shapes mixed up with finely ground clay like some indigestible pudding. It is a mucky, unappealing rock; on the map it will be marked as glacial *till* or, more particularly, boulder clay. One kind of moraine comes in swarms: these are drumlins, and because they dominate landscapes they deserve special mention. Drumlins are rounded, hummocky features usually less than a mile long: they could pass for enormous burial mounds but for their profusion. They are found in their hundreds in Wigtownshire, and in counties Down and Antrim in Northern Ireland, close around Belfast. Arthur Holmes described the scenery as 'basket of eggs', and if eggs can be twenty metres high or so, this is a good description of a collection of little rounded hillocks. They also often provide the best agricultural land; the drumlins are criss-crossed by fields, while marshes or copses dominate the less fertile depressions between them.

This account of the Ice Ages is a circuitous deviation to get back to East Anglia. The record of dramatic events is less blatantly announced there, but it is more complete. There are superficial spreads of gravel which are the products of outwash from glaciers which encroached from the North: Cromer Ridge is one notable feature with a height of up to a hundred metres – almost a mountain in East Anglian terms. This ridge has dry heath covering it, where birch trees and gorse and bracken rule. Other gravelly areas are numerous in north Norfolk, and spread down the coast in patches into Suffolk, all the way to Aldeburgh. Flat heathland betrays their origin. The stonechats and linnets that perch on the gorse bushes announce it equally, as do skittering blue butterflies, or the brown carcasses of the puffball *Bovista* which lie about like smoking golf balls. The Forestry Commission has imposed square plots with Scots pine here and there in an attempt to make this wilderness commercial. But their roots are shallow, and the winds of 1987 tipped trees up on Dunwich Forest in their thousands.

*By a reversal of this process erratics betray the direction and extent of movement of vanished ice sheets, at least if the rock is a distinctive one. The Tertiary igneous rock forming the pudding-shaped island of Ailsa Craig in the Firth of Clyde is a wide traveller; Scandinavian rocks are regularly found on the East Anglian coast.

The rocks themselves can be seen – here and there – in cliffs along the coast: that shingly, endless coast that runs between Suffolk and Norfolk and round the swell of East Anglia which juts eastwards into the North Sea like an eighteenth-century bustle. Other Crags continue above the Red Crag in a monotonous series of cross-bedded, yellow sands, with gravels and clays, exposed north of Aldeburgh. There are fossil shells in places belonging to species still alive today which tell of cold conditions – even their names (*Scalaria groenlandica*) betray the icy truth. But the famous Cromer Forest Bed, which follows along the coast (from Pakefield, near Lowestoft, northwards) shows what happened in an interglacial. Parts of it are comparatively rich in the bones and teeth of mammals. There is beaver and hyaena and sabre-toothed tiger and bear and rhino and hippo and elephant – nearly all the species are extinct. Enthusiasts can find many smaller vole-sized creatures, if they sieve the soft clays and sands carefully. The 'forest' comprises stumps of trees and pieces of wood, and there is at least one bed of peat, which I have attempted to burn – it does – slowly. Some levels yield fossil seeds.

The most amazing fossils to find in East Anglia are elephant and mammoth teeth. They can drop out of the cliffs after winter storms, or they can be grubbed up by the plough. There are several different species, depending on which glacial or interglacial they derived from: they must be among the largest chronometers in the fossil record. A fully grown tooth is larger than a loaf and ribbed like a church

Pleistocene elephant excavated from Selsey in 1961. This is the straight-tusked elephant characteristic of interglacial periods. Such finds provide unequivocal evidence for the dramatic climatic shifts in the comparatively recent geological past.

organ. They are massive objects, which is not surprising when you remember that they could grind almost any vegetable matter to a twiggy mulch. They are not particularly uncommon – in fact, elephants are large enough to attract attention when they turn up in any of the many little patches of Pleistocene that have escaped subsequent erosion. Off the Suffolk coast they occasionally get dredged up in fishermen's nets. In the Harbour Inn at Blackshore, near Southwold, there is a good specimen on the window ledge of the downstairs bar. You can admire it while you enjoy a half of Adnam's ale, which is often claimed to be the best ale in the country. This pub was immersed in the great flood of 1953, and there is a mark showing how high the water came on the wall outside.

Another relic that turns up on the beaches in Suffolk is amber. Erosion of amber-bearing deposits happens out to sea. Unlike any other fossil material, amber is light enough to float, because it is fossilised coniferous resin. When it is released from its entombment in rock it eventually finds its way shorewards. Storms accelerate the process. After a gale the professional beachcomber knows to look along the highest strand line to find any pieces of amber, where it will lurk among seaweed and detergent bottles. Real amber is valuable, but a lot of the stuff that is sold in amber shops is *copal* – relatively recent pine resin with a light yellow colour, and lacking the satisfying depths of real amber. You have to be up early to beat the locals in Suffolk. Most would-be beachcombers mistake the clear yellowish form of quartz known as citrine for amber. This occurs on the beach, too, in small, rounded stones, but it is easily recognised for what it is by its great hardness, and cold 'feel' compared with the real thing. It is an attractive enough stone, especially when polished, so why bother to look further? When searching for amber among the endless flints the occasional exotic pebble will turn up – like sparkling schist; this will almost certainly be a glacial erratic which travelled from Scandinavia on the back of an ice sheet, and what could be a more exciting discovery than such tangible proof of ice long vanished?

What you get in East Anglia is the sky. Because the rock formations lie low and flat, and are poorly consolidated, there is little to make a stand against the forces of erosion. Low relief is usual. The visitor soon becomes a connoisseur of the gentle swell or the unexpected hill. But because the land seems to crouch down out of the way – like a rabbit trying to make itself inconspicuous – your eyes flip upwards to the sky. There somehow seems to be *more* of it here.

The best days are those with a brisk wind, and mountains of grey cumulus

East Anglia. The geology provides little relief, and the sky comes to dominate the landscape, as was appreciated by artists who worked there.

breaking up and coalescing once again, and driving across the sky in every pattern and permutation inadequately described by the colour grey. Or when there are two strata of cloud: high cirrus and lower banks and powder puffs of white wooliness, and patches of clear blue to set it all against, so that you can stand and marvel at kinetics as the lower clouds skedaddle past the upper. Or you can see writ large the abstractions of the weather chart: fronts really do look like aerial walls with boundaries scribed across the sky, sun to one side, rain to the other. On occasion, there is a clear sky out to sea at sunset, unnaturally light, while the brooding clouds of an advancing front obscure the land in premature darkness; this, one imagines, is how the sky might look as the world ends.

The East Anglian artists of the 1800s are often described as the Norwich School and they appreciated the Norfolk sky. The Cromes (elder and younger), John Thirtle, James Stark and John Sell Cotman (in at least some of his work) were poets of the sky, and to see a good collection of their work is reason enough for visiting the Norwich Castle Museum. They loved blustery days, and at their best, in their sketches, they knew how to drive paint in a flourish to both imitate and capture the wind. Grey days predominate in these pictures. The geology is precisely the thing which fails to appear in most of these paintings, even though it underpins the wide horizons in these low lands. James Stark offered us views of Cromer cliffs or sand dunes as a proxy for mountains, but it is the sky we remember.

Much of the land has been drained within historical times. The fens are small remnants of what they were, and many of them are under conservation or protection orders as a result of their specially adapted species of plants and birds and insects. Somebody, fortunately, is looking out for the Great Raft Spider and the Fen Violet. Even so, these fens are little relict, 'upside-down islands' of habitat where there used to be a continent: once, wanted men could disappear into the fens and not be found for years. There is geology in the making in the fens, because it is possible to see here how peat forms from the local sedges and reeds. An augur, twisted down into the fen, records climatic and vegetation changes since the cold of the Pleistocene receded. This is a part of the country where fossils of the future are being entrapped in sedimentary rocks of the future.

The closest match for the drained part of East Anglia is probably the polder country of Holland. Even today, Dutchmen feel comfortable in fenland: I am told they are prominent among the new landowners in Suffolk and Cambridgeshire. The Dutch landscape and seascape artists have much in common with the Norwich painters, too, not least a fixation on boats, and sky. More obviously, there are Dutch gables in every East Anglian town near the sea, elegant, or quaint according to your taste. In Holland, too, the rocks are young, and were it not for man's drainage the passage from sediment to half consolidated rock and thence to mudstone would have been happening even now beneath what is currently productive polder.

Drained land makes wonderfully generous fields, flat as a snooker table, and, with the application of the appropriate fertilisers, about the same colour green. The peaty subsoil ploughs in with the surface loam to give the nearest thing to potting compost. In many sites, stones and similar irritations are rare; nobody can blame the farmer for leaping avidly upon such land. Sugar beet grows in this soil, with big, flat succulent leaves suggesting richness, but the beets themselves are rather plump and disappointingly pallid like rugby players' bottoms. The soil demands feeding, of course: it is merely the medium. In dry weather, and East Anglia is the driest part of our islands, these leafy crops demand watering as well. Such is the thirst of potatoes that great watering machines trundle up and down in summer. So this drained soil is farming factory country, mechanised and efficient and manned by the minimum number of skilled labourers with the maximum number of machines. One might say that the geology connives with all this, but time is wreaking its revenge. The peat is shrinking, and so in many places the land is sinking. Elsewhere, the land is literally blowing away. In dry weather the peat turns into light, dry granules that can be whipped into a dust storm by a strong wind. There are few hedges in such ranchland to stop this deflation. Openness results in little drifts of brown peat which run across the roads after a gale. If you run your fingers through one of these dust drifts you will discover how light it is – for the peat that serves to aeriate the soil also serves to render it vulnerable. It is too easy to wag a finger at the farmer and deliver homilies on the virtue of hedges as windbreaks in areas where the land sits upon geologically flimsy peat; it is more important to try and save what remains.

This corner of England is the most vulnerable to erosion in other ways, too. The geological infancy of the rocks means that they have not had time to become hard and resistant to erosion. They crumble. When the rocks are exposed at the coast the sea soon attacks them. Those soft sands of Pliocene to Pleistocene age along the Suffolk coast are little more than stacked dunes. Even a high wind will pick out sandgrains and whip them away. A small sandstorm may be the result: this will cause further damage by beating at the cliffs like a blaster. A winter storm can do worse. One really bad storm can wash away hundreds of tons from the cliffs. They become undermined, developing crumbling overhangs. Projecting rims of clifftop are improbably held only by the black rhizomes of bracken, which slowly rot until they suddenly discard their burden. Hence the last rocks to have accumulated in our long geological history are among the first to be destroyed.

At Easton Bavents (Suffolk) there are sandy cliffs where it is an easy matter to see the progress of destruction. A line of cottages is set at right angles to the sea at the clifftop. The track past them stops abruptly at the cliff edge. It is obvious that it once went further. Look closer, and you will see the gas pipe sticking out of the clifftop, and more besides: cables and a fragment of fence, crazily teetering. So a house must have fallen over, and recently too, judging by what remains.

The eastern part of Suffolk is fighting with the sea and losing. Erosion of the cliffs is accompanied by the relentless southward drift of shingle. The town of

Dunwich is now but a fragment. When Henry II ruled, Dunwich was the chief port of southern England. Its end was signalled by a great storm in January 1328 which blocked the harbour. Nineteen years later another storm destroyed four hundred houses in a night. Since then, the loss of land to the sea has been relentless, until today only a single street remains. There were many churches in medieval Dunwich, now all tumbled into the sea. From time to time bones are still washed ashore. The last church of the old town was All Saints, and the last service in the last church was held in 1755, before it, too, began to succumb. In 1990, the gravestone of John Easey – remaining alone in All Saints churchyard – tumbled off the clifftop, and with him the traces of what had once been a great and populous town. It is still possible to see the abbey walls, and an improbably grand arch leading nowhere. There is a model of the town in its heyday in a small museum occupying one of the last remaining cottages.

So, too, at Aldeburgh to the south. Here, the destruction has not proceeded so far. But the oak-beamed and ancient Moot Hall, which must have been at the centre of the town, now abuts the shingle bar. Winter storms fling flint pebbles against it. A comparatively new sea front gives Aldeburgh a cocksure air, but the waves will win in the end, and the Brudenell Hotel, and the smart but slightly faded holiday homes will go with it into the sea that has already swallowed Dunwich. Only flint endures.

The march of pebbles has constructed Orford Ness southwards from Aldeburgh like a long, crooked finger. There are good records of how it has changed in historical times. Every year its outline is a little different. The river Alde once reached the sea at Aldeburgh, but as the Ness grew its course was deflected ever further south, until now its deviation runs to a dozen miles. New land is, in a sense, a compensation for the loss elsewhere, although there is precious little that can be done with a great pile of shingle. Only that most mysterious of breeds, the sea fisherman, visits it regularly, and may be seen sitting apparently motionless beneath a huge green umbrella. On a bitter winter day it would be hard to beat this shingle bar for desolation; intractable and exposed. The Ness is a place of geology in the making – a cycle of construction and destruction, and nobody knows how it will balance out in the end. A small sea level rise as the result of the greenhouse effect might be disastrous, for example. So this piece of East Anglia is a testing ground for Man's follies; let us hope that we never regret siting the nuclear power station Sizewell B here on such an insubstantial shore.

Above   *Unconsolidated sands make up the cliffs at Cromer, Norfolk. Coastal erosion is very rapid in many parts of East Anglia.* Below left   *Peat shrinkage, exposing tree roots in Leziate Fen, Ashwicken, Norfolk; over only 8 years the surface of the peat fell 20 cm (8 in).* Below right   *The reed* Phragmites *grows extensively around the Broads, and behind coastal shingle bars, and provides the habitat for ornithologist's specialities, bearded reedlings and marsh harriers.*

Nearer Dunwich, the shingle bar runs like a rampart between sea and land. Behind it, the bird reserve at Minsmere has been created with a combination of nature and artifice. The latter was provided by Mr Axell, who managed reeds and mere, scrape and islands in a way that improved on nature only by understanding it. He is a hero of the conservation movement, whose lack of stridency only adds to his achievement. This is where the avocet returned to breed again after an absence of a century. There is a fine exposure of cross-bedded sands in the new reserve car park. Sand martins appreciate the exposure, too, for they excavate their nest burrows in the unconsolidated rocks. Dozens of them zoom back and forth over the parked cars in the summer, feasting on insects while the bird watchers eat their corned beef sandwiches far below.

Northwards again, the Westwood marshes near Walberswick are the largest unbroken stretch of *Phragmites* reeds in the country, where the marsh harrier flaps, lazily oblivious of squads of binoculars trained upon him. This must be the only part of the country where eighty per cent of the casual ramblers carry binoculars, and where 'Anything about?' is a regular form of greeting. The shingle rampart forms the horizon once again. On a spring day, while you listen to the boom of the bittern it is easy to forget the fragility of this coast, and how a dozen mighty waves could smite it to extinction.

All around the East Anglian coast there are protected bays and estuaries with salt marshes. This is another site where rocks are being made, for the silt brought down by full rivers builds up the marsh slowly, while it is also drowned beneath an exceptional high tide once in a while. It is an in-between kind of place, too salty for all but specialised plants. Most arresting of these is the sea lavender (*Limonium*) which can colour patches yards across with the lilac colour of the Highlands, although it is no relative of the heath family. Glaucous plants are typical, like the sea purslane (*Halimione*). The samphire also grows there, leafless and jointed like a miniature cactus, but springing directly from the mud. It is curious how having to cope with salt produces succulent plants, yet the samphire is immersed in water for half its life. Richard Mabey recommends it as a vegetable, and indeed it is perfectly palatable cooked with butter, with a salty-crunchy taste, but I am not sure I am gourmet enough to prefer it to asparagus. There are clams, like cockles, that occur in great numbers in the mudflats adjacent to saltmarsh, and the green crab is abundant there. If you brave the slithery wastes when the tide is out it is a wonderful place to see how the fossils of tomorrow are being preserved today.

Another specialist area of East Anglia is the Breckland, where the layer of sands overlying chalk and the almost arid climate conspire to create special conditions. This being the driest part of Britain, and the soil extremely light, the typical plants of the brecks are annuals. They germinate quickly, proceed rapidly from germination to flower, and then fade as fast. This is the same strategy as that used by desert annuals, utilising resources when conditions are favourable after rain, when the desert 'blooms'. Sadly, most of the Breckland specialities are not that spectacular, although that does not prevent the ardent botanist from cooing with delight when discovering one for the first time. Several of the Breckland speedwells are exceptionally modest plants, and best seen by crawling around with your nose a few centimetres from the ground. The spiked speedwell (*Veronica spicata*) is a more aesthetically attractive plant altogether, with its bright blue bottle brush flowers – so much so that it is protected by law. The army has reserved great tracts of this natural heathland for its own use, and afforestation accounts for more, and as a result the Breckland flora is not easy to find. The annuals require the soil to be disturbed to stimulate germination of their seeds. Presumably the army does this whenever tanks churn up the ground, or at target practice. In nature reserves this disturbance has to be managed (rabbits make their contribution, too). None of this creates enough habitat to keep up the population of stone curlews, a bird which will decline like the corncrake. It is all a most delicate enterprise, this preservation of an ecosystem. We find it difficult to manage limited conservation in one corner of our small island, even with the support of an army of volunteer labourers and a phalanx of twitchers. It is as well to remember this when we adopt a high moral tone with those struggling to preserve thousands of square miles of Brazilian rain forest with little in the way of resources.

There are clays dotted over parts of East Anglia, of Cretaceous age in the west, of Pleistocene or younger age in the east, which have provided the local bricks which lend appeal to many ordinary buildings. In Chalkland villages these are combined with flints, as was described previously. Towards the coast, warm red colours are dominant, which can make an attractive pairing with flint. The local character of many eastern bricks is revealed by the burnt flints that they may contain. Sadly, some of them are not too tough either, which means that they have to be replaced by mass produced substitutes that are often a poor colour match. There are some extraordinary double sized bricks in Eastern Suffolk, which are worth looking out for. We are so used to seeing, as it were, brick-sized bricks that these big ones look unnatural. They were presumably inspired by the simple notion that if you made bricks twice the size you would need half as many to build a house. The result is that you have to look twice to reassure yourself that what you see is actually brickwork and not some curious species of local stone. They do not appear to be especially durable and may not endure, which is a pity, because every local curiosity adds to the richness of the rural scene.

Where there are no thatched roofs there are pantiles. They were imported in the seventeenth century from the Netherlands as a durable alternative to thatch, and they are also lighter on the roof beams than ordinary tiles, requiring fewer overlapping courses. Some, at least, were made from local clays in the eighteenth and nineteenth centuries. Although the rather coarse texture of pantiles is popularly associated with the Mediterranean, they have an authentically East Anglian to Midlands flavour in British vernacular architecture. I particularly like the black ones – rarer than red – and even more when they are combined with tar-washed sea cottages on the Norfolk coast. These are the darkest houses there are: the owners should be required, by law, to wear eye patches and chew plug tobacco.

Swifts seem to favour pantiled roofs. Where the tiles meet the edge of the roof there may be a hole, nicely swift-sized, from which the birds can launch directly into the air in pursuit of their insect prey. They make a great mess under the roof – a matted pile of guano, feathers and straw. When the east winds blow in the winter, as they often do, snow drives in through the swift holes. Thatch may, after all, suit better the harsher aspects of the East Anglian climate.

The Norfolk Broads are as artificial as are the dry fields of Fenland. They look 'natural', and their use for boating holidays has made them a cherishable national asset. But they originated as peat diggings. They are just as man-made as the flooded gravel pits that now provide accommodation for small marinas along the Thames Valley. The Broads are the product of a great peat industry in the thirteenth and fourteenth centuries, when open cast mining for fuel was conducted on a huge scale (200,000 turves per annum from South Walsham alone according to contemporary records). The accumulation of peat was finite, as it always is, and when the reserves were nearly exhausted flooding of the pits happened in places to a depth just right for colonisation by reeds; which were

Facing page   *Man's mark is left upon the landscape. Fossil footprints were left by Palaeolithic Man upon muds in the Usk estuary, and have now been exhumed by the vagaries of erosion.*

then followed by a succession of other plants as silt was trapped by the stems. When eventually alder carr becomes established the broad is on its way towards becoming land again. It is a strange paradox that mining of post-glacial peat has made an artificial landscape which inspires loyalty and a conservationist attitude denied the natural heathland elsewhere in East Anglia. Yesterday's wilderness is tomorrow's national park.

East Anglia shows the interplay between construction and destruction, between sediment and erosion, between ice and warmth. These cycles may yet continue.

We are now in an interglacial. It has not lasted as long as several interglacials before. How do we know that the great ice sheets will not once more sweep down from the North, maybe polishing and scouring the fragile rock records of previous Ice Ages, so that it will be as if they had never been; or perhaps adding just one more boulder clay to the complex sequence on the Norfolk coast? Mankind's artefacts may be preserved as one more set of traces to be dumped on top of the flint handaxes, and flaked scrapers of our precursors – just another entry into the vast ledger of the history of the British Isles, added to its most vulnerable corner. We could prove but a brief interlude, no thicker than a bed of peat in the real chronology in which the history of the Earth is calibrated. Some have claimed that the ice could return even within a few years as a kind of freezing catastrophe. Others believe that we are out of this glacial phase forever. I have heard contrary accounts of what the 'greenhouse effect' and the holes in the ozone layer could do to the whole equation; capacious computers are refining models to try and inspire confidence that humankind is somehow in control. The geological history of our islands does not inspire confidence that this will be the case. But buffeted by chance, we may yet survive.

Run sand scooped from the soft cliffs of Norfolk through your hands: as in an hourglass, this is ephemeral time. Will humankind endure longer than these cliffs, which crumble daily into the indifferent sea?

Glossary

I have introduced a modest number of terms in this book some of which may be unfamiliar to the reader. It will be helpful to have a short glossary to summarise these words.

agglomerates – deposits formed by coarse fragments of volcanic rocks, especially associated with the 'necks' of ancient volcanoes.

anticline – structure in which the strata are bowed, or folded upwards; the dip of the beds is away from the centre of the anticline.

basalt – black, fine grained volcanic rock, the typical extrusions from the volcanoes of oceanic islands.

batholith – a major intrusion of igneous rock – such as granite or gabbro – often associated with phases of mountain building.

bed – an individual layer or stratum of sedimentary rock.

boulder clay – a rock associated with Ice Age deposition, typically a heterogeneous mixture of clay, and pebbles and boulders of various shapes.

calcite – the crystalline form of calcium carbonate used in the construction of shells of many animals which are found as fossils.

calcareous – descriptive of limy, or calcitic, rocks and soils.

carnosaurs – meat-eating dinosaurs.

chert – the general term for flint-like rocks, composed of silica in its glass-like form.

clay-with-flints – superficial deposit overlying the Chalk, named after its typical constituents.

cleavage – the direction in which a slate, or a mineral splits.

conglomerates – coarse-grained rocks composed of more or less rounded cobbles or boulders of other rocks, often cemented together with sandstone or quartzite.

cross bedding – sets of bedding planes inclined at an angle, reflecting deposition by currents, or by the action of wind.

dolomite – a common rock forming mineral which is a carbonate of calcium and magnesium.

evaporites – deposits formed from the evaporation of former oceans, typically including rock salt (halite).

facies – describing the local type of sedimentary deposition.

foliation – the banding in a metamorphic rock produced by the preferred alignment of certain minerals.

foraminiferan – a small, single celled organism with a calcareous shell or test.

gabbro – a dark, coarse-grained igneous rock.

genus – rank of classification of animals and plants that usually includes several, similar species. *Rosa* or *Fuchsia* are familiar, garden examples.

gneiss – a common metamorphic rock, usually rather coarsely pink-and-black banded.

granite – the general term for coarse-grained, igneous intrusive rocks composed predominantly of the minerals quartz, felspar and mica.

graptolite – extinct, colonial animals that were part of the Lower Palaeozoic plankton, leaving hacksaw blade-like fossils in shales.

grit – a coarse-grained and rough weathering sandstone.

halite – rock salt, sodium chloride, an important evaporite.

Hercynian – applied to the phase of mountain building which folded the Carboniferous and Devonian rocks of the West Country.

igneous – 'fire formed' – those rocks which solidified from liquid magma including both volcanic rocks and plutonic ones, like granite.

induration – the hardening of soft sediments by natural processes, particularly by the introduction of cements like calcite.

intrusion – body of igneous rock, which may be very large, introduced often as magma into the country rock.

Laurentia – a term applied to the ancient North American continent, including Greenland, and North-west Scotland (Chapter 4).

magma – 'liquid rock' – the melt originating deep in the Earth's crust from which igneous rocks are formed.

Mesozoic – Trias, Jurassic and Cretaceous – the time during which the largest land animals were reptiles.

metamorphism – the process of changing the character of rocks by heat and pressure, often far below the surface during phases of orogeny.

migmatite – a metamorphic rock of high grade, including seams of granite-like rock.

mudtonstone – a rock formed of consolidated clay, but without the well developed fine bedding of a typical shale.

mylonite – a fine grained rock produced by the enormous grinding pressure beneath a thrust (e.g. below the Moine Thrust).

nappes – huge structures developed in mountain belts displacing masses of rock over enormous distances.

oolite – rock composed of spherical grains (ooids), usually composed of limestone, and forming some of the most important building-stones.

orogeny – phase of mountain building, often caused by the approach of two lithospheric plates.

Palaeozoic – literally 'old life', that period of time embracing the Cambrian to Permian; divided at the end of the Silurian into a Lower and Upper Palaeozoic.

petrology – the study of the mineral composition and fabric of rocks.

Precambrian – that vast stretch of time prior to the Cambrian Period.

quartz – the common crystalline form of silica (silicon dioxide), including its pure, colourless variety 'rock crystal'.

schist – finely foliated metamorphic rock, usually with mica, found over great areas of the Highlands.

sedimentary – describing rocks that were laid down as sediments, by water, wind or ice.

shales – sedimentary rocks made of mud or silt, which split into sheets along the bedding planes; slates are shales which have been hardened in the early phases of metamorphism, and which cleave along planes which are not necessarily related to the original bedding.

sill – a body of igneous rock which has been concordantly emplaced between beds of sedimentary rock, thus forming a planar sheet.

silica – silicon dioxide, forming quartz and flint, and one of the most abundant materials in the Earth; the majority of rock-forming minerals are silicates.

subduction – the process whereby, during the convergence of two lithospheric plates, basalt oceanic crust is consumed down a subduction zone, which becomes a centre for earthquake and volcanic activity.

syncline – a downward fold such that the beds dip towards the centre, where the youngest beds are found.

terrestrial – formed or living on land, as opposed to subaqueous.

till – the general term for the deposits of glaciers and ice.

unconformity – a contact between two rock formations (often recorded in a change in the angle of the dip of the beds), representing a considerable break in continuity, and time.

Appendix – The Geological Map

Geological maps show a graphic picture of the rocks and the superficial deposits at or near the surface. These maps are produced in distinctive colours which are generally printed over a "subdued" topographic base to provide the geographical details needed to locate oneself on the ground. Precise positions may be determined by reference to the lines of the National Grid.

Geological maps are often illustrated with two-dimensional cross sections, showing what the rock structure actually looks like beneath the surface at certain locations. The map will also identify the relative ages of the rocks themselves and denote, by way of a map key, the order in which they were formed. A geological map of a location forms an important ingredient in the understanding of its geology, natural resources, soils and the landscape.

What geological maps are available?

A wide range of geological maps at different scales, are available from the British Geological Survey. Our graphic index map illustrated overleaf, shows the availability of Geological Maps at the popular scales of 1:63,360 (one-inch) and 1:50,000. This illustration, and the listing of the map sheets, has been extracted from a Catalogue of Printed Maps and Associated Literature, which is available from BGS, Keyworth, Nottingham, NG12 5GG (Tel 0602 363100).

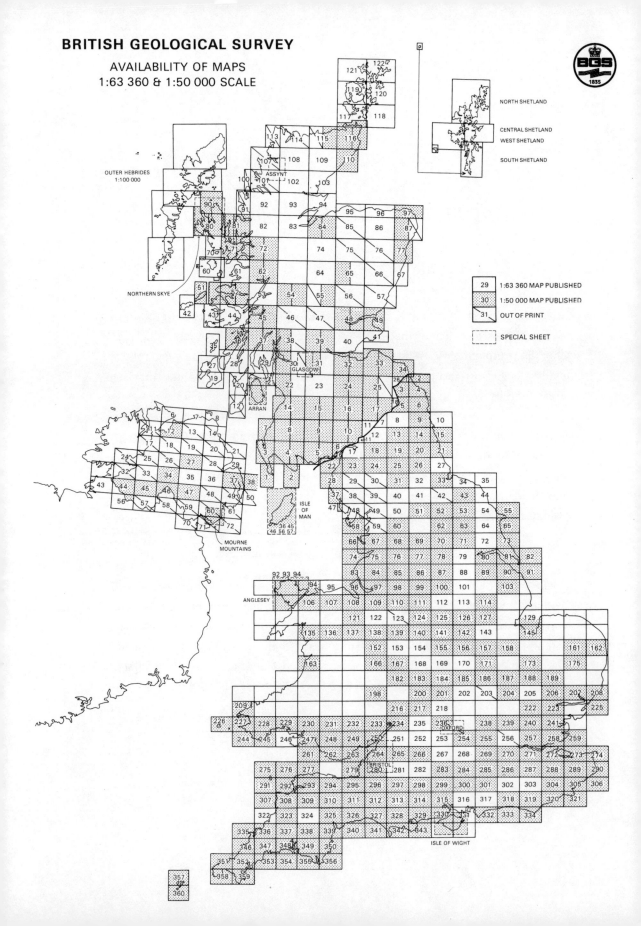

BRITISH GEOLOGICAL SURVEY

AVAILABILITY OF MAPS
1:63 360 & 1:50 000 SCALE

29	1:63 360 MAP PUBLISHED
30	1:50 000 MAP PUBLISHED
31	OUT OF PRINT
	SPECIAL SHEET

NORTH SHETLAND

CENTRAL SHETLAND

WEST SHETLAND

SOUTH SHETLAND

OUTER HEBRIDES
1:100 000

ASSYNT

NORTHERN SKYE

ARRAN

ISLE
OF
MAN

MOURNE
MOUNTAINS

ANGLESEY

GLASGOW

OXFORD

BRISTOL

ISLE OF WIGHT

1:63 360

England and Wales

4	Holy Island (D) 1925
7	Kielder Castle (S) (D) 1950
8	Elsdon (S) (D) 1951
9	Rothbury (S) 1934
10	Newbiggin (S) (D) 1934
11	Longtown (D) 1925
12	Bewcastle (S) (D) 1969
14	Morpeth (S) 1955
15	Tynemouth (D) 1968
16	Silloth (D) 1925
17	Carlisle (D) 1925
22	Maryport (S) 1930
27	Durham (S) (D) 1965
34	Guisborough (D) Pre-1900
34 & 44	Whitby & Scalby (D) Pre-1900
40	Kirkby Stephen (S) (D) 1972
41	Richmond (S) (D) 1970
43	Egton (D) Pre-1900
44	(see 35)
50	Hawes (S) (D) 1971
52	Thirsk (D) Pre-1900
54	Scarborough (D) Pre-1900
64	Great Driffield (C) Pre-1900
68	Clitheroe (S) 1960
72	Beverley (D) Pre-1900
75	Preston (D) 1940
76	Rochdale (S) 1927
79	Goole (S) 1972 (D) 1971
84	Wigan (D) 1935
85	Manchester (D) 1930
88	Doncaster (S) (D) 1969
95	Rhyl (S, D) 1970
98	Stockport (D) 1962
101	East Retford (C) 1967
108	Flint (D) 1924
110	Macclesfield (D) 1968
112	Chesterfield (C) 1963
113	Ollerton (C) 1966
122	Nantwich (S) (D) 1967
143	Bourne (S) (D) 1964
152	Shrewsbury (D) 1932
153	Wolverhampton (S) (D) 1929
154	Lichfield (S) 1926 (D) 1922
166	Church Stretton (D) 1967
168	Birmingham (S) (D) 1924
169	Coventry (S) 1926 (D) 1922
170	Market Harborough

	(C) 1969
202	Towcester (C) 1969
205	Saffron Walden (D) 1952
218	Chipping Norton (C) 1968
229	Carmarthen (D) 1967
235	Cirencester (D) 1933
246	Worms Head (D) 1960
247	Swansea (D) 1972
248	Pontypridd (S) 1963
249	Newport (D) 1969
251	Malmesbury (C) 1970
252	Swindon (C) 1974
253	Abingdon (S) (D) 1971
267	Hungerford (D) Pre-1900
268	Reading (D) 1904
281	Frome (C) 1965
282	Devizes (D) 1959
302	Horsham (C) 1972
303	Tunbridge Wells (C) 1971
316	Fareham (D) 1933
317	Chichester (D) 1957
322	Boscastle (C) 1969
324	Okehampton (C) 1969
331	Portsmouth (C) 1964

Special Sheet
Bristol district (C) 1962

Scotland

11	Langholm (S) (D) 1968
19	Bowmore (C) Pre-1900
22	Kilmarnock (S) 1928
23	Hamilton (S) (D) 1929
26	Berwick upon Tweed (C) 1969
27	Port Askaig (C) 1900
28	Jura (C) 1911
29	Rothesay (C) 1971
30	Glasgow (S) 1958 (D) 1961
31	Airdrie (S) (D) 1924
32	Edinburgh (D) 1967
35	Colonsay (C) 1911
37	Inveraray (S) 1903 (D) 1908
40	Kinross (S) 1971 (D) 1973
41	North Berwick (S) 1930 (D) 1971
42 & 50	Tiree (C) 1926
43	Iona (D) 1954
44	Mull (S) 1923 (D) 1954
45	Oban (S) (D) 1926
52	Tobermory (D) 1968

53	Ben Nevis (C) 1948
60	Rhum (C) 1971
61	Arisaig (S) 1971 (D) 1969
64	Kingussie (C) 1913
65	Balmoral (C) 1904
67	Stonehaven (C) 1929
70	Minginish (D) 1913
74	Grantown on Spey (C) 1914
82	Lochcarron (S) (D) 1913
83	Inverness (S) 1914 (D) 1954
84	Nairn (S) 1958
86	Huntly (D) 1923
92	Inverbroom (C) 1913
93	Alness (C) 1912
94	Cromarty (S) 1973 (D) 1972
95	Elgin (S) (D) 1969
96	Banff (S) 1955 (D) 1954
102	Lairg (C) 1925
103	Golspie (S) (D) 1950
108	Altnaharra (C) 1931
109	Achentoul (C) 1931
110	Latheron (C) 1913
116	Wick (D) 1913
117	Hoy (D) 1932
118	Copinsay (D) 1932
119	Kirkwall (D) 1932
120	Stronsay (D) 1932
121	Westray (D) 1932
122	Sanday (D) 1932

Special sheets
Assynt (S with some D) 1965
Northern Shetland (S) (D) 1968
Western Shetland (S) (D) 1971
Central Shetland (S) 1982 (D) 1981
Southern Shetland (S) (D) 1978

Northern Ireland

7	Giant's Causeway (S) (D) 1968
8	Ballycastle (S) (D) 1963
29	Carrickfergus (S) (D) 1968
35	Dungannon (S) (D) 1961
36	Belfast (S) (D) 1966

1:50 000

England and Wales
1 & 2	Berwick-on-Tweed & Norham (S, D) 1977
3	Ford (S) 1976; (D)

	1979
4	Holy Island (S) 1974
5	The Cheviot (C) 1976
6	Alnwick (S) (D) 1972
9	Rothbury (D) 1977
13	Bellingham (S) (D) 1980
14	Morpeth (D) 1977
15	Tynemouth (S) 1975
18	Brampton (S) 1976 (D) 1980
19	Hexham (S) 1975
20	Newcastle upon Tyne (S) 1989
21	Sunderland (S, D) 1978
22	Maryport (D) 1980
23	Cockermouth (S) 1977 (D) 1975
24	Penrith (S) (D) 1974
25	Alston (C) 1973
26	Wolsingham (S) (D) 1977
28	Whitehaven (S) 1979 (D) 1976
31	Brough-under-Stainmore (S) (D) 1974
33	Stockton (C) 1987
37 & 47	Gosforth and Bootle (S, D) 1980
51	Masham (S) (D) 1985
53	Pickering (D) 1973
55 & 65	Flamborough and Bridlington (S, D) 1986
58	Barrow-in-Furness (S, D) 1976
60	Settle (S) 1989
62	Harrogate (S) (D) 1987
63	York (C) 1983
65	(see 55)
66	Blackpool (S, D) 1975
67	Garstang (S) 1990 (D) 1991
68	Clitheroe (D) 1975
69	Bradford (S) (D) 1974
70	Leeds (S) (D) 1974
71	Selby (S) 1973
73	Hornsea (D) 1909 (1990F)
74	Southport (S, D) 1989
75	Preston (S) 1982
76	Rochdale (D) 1974
77	Huddersfield (C) 1978
78	Wakefield (C) 1978
80	Kingston upon Hull (S) (D) 1983
81 & part 82 & 90	Patrington (C) 1991
83	Formby (S) 1976

(D) 1974

84 Wigan (S) 1977
85 Manchester (S) 1975
86 Glossop (C) 1981
87 Barnsley (C) 1976
89 Brigg (S) (D) 1982
90 & 91 Grimsby (C) 1990
94 Llandudno (S, D) 1989
96 Liverpool (S) 1974 (D) 1975
97 Runcorn (S) 1980 (D) 1977
98 Stockport (S) 1976
99 Chapel en le Frith (S) 1975 (D) 1977
100 Sheffield (C) 1974
103 Louth (C) 1980
106 Bangor (S) 1985
107 Denbigh (S) (D) 1985
108 Flint (S) 1972
109 Chester (S) 1986; (D) 1965 (1990F)
110 Macclesfield (S) 1968 (1990F)
111 Buxton (S) (D) 1978
114 Lincoln (C) 1973
121 Wrexham (S) 1973 (D) 1927 (1990F)
124 Ashbourne (C) 1983
125 Derby (C) 1972
126 Nottingham (C) 1972
127 Grantham (C) 1972
135 Harlech (S) (D) 1982
136 Bala (S) 1986
137 Oswestry (S) 1972 (D) 1975
138 Wem (C) 1967 (1990F)
139 Stafford (D) 1974
140 Burton upon Trent (C) 1982
141 Loughborough (C) 1976
142 Melton Mowbray (D) 1977
145, part 129 King's Lynn and The Wash (C)
152 Shrewsbury (S) 1978
155 Coalville (C) 1982
156 Leicester (C) 1975
157 Stamford (C) 1978
158 Peterborough (C) 1985
161 Norwich (C) 1975
162 Great Yarmouth (C) 1990†
163 Aberystwyth (S) 1984 (D) 1989
166 Church Stretton (S) 1974
167 Dudley (C) 1975
171 Kettering (D) 1976
173 Ely (C) 1980
175 Diss (C) 1989
182 Droitwich (C) 1976

183 Redditch (C) 1989
184 Warwick (C) 1985
185 Northampton (C) 1980
186 Wellingborough (C) 1974
187 Huntingdon (D) 1975
188 Cambridge (C) 1981
189 Bury St Edmunds (C) 1982
198 Hereford (C) 1989
200 Stratford-upon-Avon (C) 1974
201 Banbury (C) 1982
204 Biggleswade (D) 1976
206 Sudbury (C) 1991
207 Ipswich (C) prov. ed. 1990
208 & 225 Woodbridge & Felixstowe (C) 1977
216 Tewkesbury (C) 1988
217 Moreton-in-Marsh (C) 1981
222 Great Dunmow (C) 1990
223 Braintree (C) 1982
226 & 227 Milford (S) (D) 1978
228 Haverfordwest (S) (D) 1976
229 Carmarthen (S) 1975
230 Ammanford (S) (D) 1977
231 Merthyr Tydfil (S) (D) 1979
232 Abergavenny (C) 1990
233 Monmouth (C) 1974
234 Gloucester (C) 1972
236 Witney (C) 1982
238 Aylesbury (D) 1923 (1990F)
239 Hertford (D) 1978
240 Epping (C) 1981
241 Chelmsford (C) 1975
244 & 245 Pembroke and Linney Head (C) 1983
247 Swansea (S) 1977
248 Pontypridd (S) 1975
249 Newport (S) 1975
250 Chepstow (C) 1972
254 Henley-on-Thames (C) 1980
255 Beaconsfield (D) 1974
257 Romford (D) 1976
258 & 259 Southend & Foulness (C) 1976
261 & 262 Bridgend (C) 1990
263 Cardiff (S) 1986 (D) 1989
264 Bristol (C) 1974
265 Bath (C) 1965 (1990F)
266 Marlborough (D) 1974
269 Windsor (C) 1981
270 South London (C)

1981
271 Dartford (D) 1977
272 Chatham (D) 1977
273 Faversham (C) 1974
274 Ramsgate (S, D) 1980
277 Ilfracombe (C) 1981
279 Weston-super-Mare (C) 1980
280 Wells (C) 1984
283 Andover (D) 1975
284 Basingstoke (C) 1981
285 Aldershot (D) 1976
287 Sevenoaks (C) 1971 (1990F)
286 Reigate (D) 1978 (C) 1980
288 Maidstone (C) 1976
289 Canterbury (C) 1982
290 Dover (C) 1977
292, parts 275, 276, 291 & 308 Bideford & Lundy (C) 1977
293 Barnstaple (C) 1982
294 Dulverton (C) 1974
295 Taunton (C) 1984
296 Glastonbury (C) 1973
297 Wincanton (C) 1972
298 Salisbury (D) 1976
299 Winchester (D) 1976
300 Arlesford (D) 1975
301 Haslemere (C) 1981
304 Tenterden (C) 1981
305 & 306 Folkestone & Dover (C) 1974
307 & 308 Bude (C) 1980
309 Chulmleigh (C) 1980
310 Tiverton (C) 1974
311 Wellington (D) 1976
312 Yeovil (C) 1973
313 Shaftesbury (D) 1977
314 Ringwood (D) 1976
315 Southampton (C) 1987
318 & 333 Brighton and Worthing (C) 1984
319 Lewes (C) 1979
320 & 321 Hastings and Dungeness (C) 1980
323 Holsworthy (C) 1974
325 Exeter (D) 1971 (1986F)
326 & 340 Sidmouth (D) 1974
327 Bridport (S) 1977 (D) 1974
328 Dorchester (D) 1981
329 Bournemouth (C) 1991
330 Lymington (D) 1975
332 Bognor (D) 1975
333 (see 318)
334 Eastbourne (C) 1979
335 & 336 Trevose Head & Camelford (C) 1976
337 Tavistock (D) 1977
338 Dartmoor Forest (D) 1977
339 Newton Abbot (C) 1976
340 (see 326)

341 & 342 West Fleet & Weymouth (D) 1976
343, part 342 Swanage (C) 1976
346 Newquay (D) 1981
347 Bodmin (D) 1982
348 Plymouth (D) 1977
349 Ivybridge (D) 1974
350 Torquay (D) 1976
351 & 358 Penzance (C) 1984
352 Falmouth (C) 1990
353 & 354 Mevagissey (C) 1975
355 & 356 Kingsbridge & Start Point (D) 1975
357 & 360 Isles of Scilly (C) 1975
359 Lizard (D) 1975

Special sheets
Isle of Man (D) 1975
Isle of Wight (D) 1976
Anglesey (D) 1974 (S) 1980

Scotland

1 Kirkmaiden (S) 1985 (D) 1982
2 Whithorn (S, D) 1987
3 Stranraer (S) (D) 1982
4W Kirkcowan (S) 1978 (D) 1982
4E Wigtown (S) 1978 (D) 1981
5W Kirkcudbright (S) 1977 (D) 1980
5E Dalbeattie (S) 1977 (D) 1981
6 Annan (D) 1980
7 Girvan (S) 1988 (D) 1980
8W Carrick (S) 1978 (D) 1981
8E Loch Doon (S) 1977 (D) 1980
9W New Galloway (S) 1978 (D) 1979
9E Thornhill (S) 1978 (D) 1980
10W Lochmaben (D) 1983
10E Ecclefechan (D) 1982
14W Ayr (S) (D) 1978
14E Cumnock (S) 1976 (D) 1980
15W New Cumnock (S) 1986 (D) 1982
15E Leadhills (S) 1987 (D) 1981
16W Moffat (D) 1987
16E Ettrick (D) 1987
17W Hawick (S) (D) 1982
17E Jedburgh (S) (D) 1982
22W Irvine (D) 1987
24W Biggar (S) 1980 (D) 1981

† includes offshore geology

24E   Peebles (S) 1979 (D) 1983
25W   Galashiels (S) 1983 (D) 1985
30W, part 29E Greenock (D) 1989 (S) 1990
32W   Livingston (S) 1977
32E   Edinburgh (S) 1977
33W, part 41 Haddington (S) 1983 (D) 1978
33E, part 41 Dunbar (S) 1980 (D) 1978
34   Eyemouth (S) 1982 (D) 1983
36   Kilmartin (D) 1974
37W   Furnace (S) 1989
37E   Lochgoilhead (S) 1990
38W   Ben Lomond (S) 1987
39W   Stirling (S) (D) 1974
39E   Alloa (S) (D) 1974
48W   Perth (S) 1983 (D) 1985
48E   Cupar (S) (D) 1982

49   Arbroath (S) 1980 (D) 1981
51W   Coll (C) 1976
51E   Caliach Point (C) 1976
52W   Ardnamurchan (S) 1977
52E   Strontian (S) 1977
54W   Blackwater (C) 1974
54E   Loch Rannoch (C) 1974
55E   Pitlochry (S) 1981
62W   Loch Quoich (S) (D) 1975
62E   Loch Lochy (S) (D) 1975
65W   Braemar (S) 1989
71W   Broadford (S) (D) 1976
71E   Kyle of Lochalsh (S) (D) 1976
72W   Kintail (S) 1984 (D) 1985

72E   Glen Affric (S) 1986
77   Aberdeen (S) 1982 (D) 1980
80W   Dunvegan (S) (D) 1975
80E, part 81 Portree (S) (D) 1975
81W   Raasay (D) 1980
81E   Loch Torridon (D) 1975
84W   Fortrose (D) 1978
84E   Nairn (D) 1978
87W   Ellon (S) 1991
90, part 91 Staffin (S) (D) 1976
97   Fraserburgh (S, D) 1987
110   Latheron (S) 1985
115E   Reay (S) 1985
116W   Thurso (S) 1985
116E   Wick (S) 1986

Special sheet
Arran (S) 1987 (D) 1985

Northern Ireland

6 & 12 Limavady (S) (D) 1981
7   Cookstown (S) (D) 1985
34   Pomeroy (S) 1979 (D) 1978
37 & 38 Newtownards (S) 1989 (D) 1988
44, 56 & 43 Derrygonnelly & Marble Arch (D) 1989 (S) 1991
45   Enniskillen (S) (D) 1982
46   Clogher (S) (D) 1983
47   Armagh (S) (D) 1985

Special sheet
Mourne Mountains (S) 1978

Acknowledgments

I am indebted to Heather Godwin for commissioning this book, and for sustaining its completion with her enthusiasm. Hamish Francis kindly corrected the proofs. The illustrations derive from several sources. The unrivalled archive of geological photographs at the British Geological Survey provided the greatest number, and I am indebted to the Director of the Survey for facilitating their use. Sylvia Brackell and Clive Jeffery were always helpful in directing me towards the appropriate sources, and obtaining transparencies and prints. Another important source of illustrations was the slide collection in The Natural History Museum. I am deeply indebted to Alan Timms and Marilyn Carter for their help in locating slides for reproduction, without which the book would have been much the poorer. My thanks also to Joyce Pope for permission to use her wonderful plant photographs. Numerous friends and colleagues contributed illustrations from their collections: particular thanks go to Philip Lane, Andrew Smith, Peter Forey, Peter Barnard, Anthony Sutcliffe, Angela Milner, Michael Day, Robert Kruszynski, Henry Buckley and Andrew Clark. Robin Cocks lent me numerous books from his personal library, and read and improved the manuscript. Martin Pulsford obtained further illustrations from the library of transparencies in the Publications Section of the Natural History Museum. Finally, my wife Jacqueline had to tolerate the ill temper of the author during a year of composition, but was still generous enough to read and improve the resulting text.

Select Bibliography

The publications of the British Geological Survey are listed above. Further books of general interest are given here. Many readers will be familiar with the classic *New Naturalist* series (Collins) which are models for a crosss-cultural approach to natural history for discrete areas of the British Isles. Although some volumes have been reprinted, it is worth exploring secondhand bookshops for the original editions. They are already collector's items.

The Geologist's Association does a splendid job on behalf of those with an amateur interest in rocks and fossils. The Association is at Burlington House, Piccadilly, London W1V 9AG. Their local guides explore the rocks of many of the areas described in this book in much more detail than I have given. These guides are all £10 or less. I can particularly recommend: The Isle of Wight (1984, £2.00); The Isle of Arran (1989, £5.00); The Geology of the Dorset Coast (1989, £7.50); The Lake District (1990, £10.00); Manchester area 1991, £9.00); Yorkshire Coast (1992, £8.00); Late Precambrian geology of the Scottish Highlands and Islands (1991, £10.00); Isle of Man (1993, £8.50). There are also two entertaining illustrated geological walks around London (£4.95 each).

The Geological Society of London at Burlington House publishes more academic texts, but three works published by the Society will give up-to-date information for real enthusiasts:

McDuff, P. & Smith, A.J. (eds) *The Geology of England and Wales*, 1992.

G.Y. Craig (ed) *Geology of Scotland*, 3rd ed., 1991.

Cope, J.C.W., Rawson, P.F. & Ingham, J.K. (eds) *Atlas of Palaeogeography and lithofacies*, Geological Society Memoir no.13, 1992. (This book is too expensive

for the average reader, but public libraries should be encouraged to purchase
it.)

The Natural History Museum, Cromwell Road, London, SW7 5BD, publishes
useful general guides to the fossils of the Palaeozoic, Mesozoic and Tertiary rocks of Britain, as separate volumes. The Palaeontological Association
publishes guides to the fossils of the Chalk, Oxford Clay (Jurassic), Permian
reefs, and the London Clay. These latter may be obtained by post from Dr L.
Cherns, Geology Department, University College, Cardiff.

General books which may help the reader broaden the appreciation of geological
dimension of history and landscape include:

Fortey, R.A. *Fossils: The Key to The Past*, 2nd ed., Natural History Museum
Publications, 1991.

Hoskins, W.G. *The Making of the English Landscape*, Hodder & Stoughton, 1955;
Penguin Books, 1985.

Oldroyd, D.R. *The Highlands Controversy*, University of Chicago Press, 1990.

Rackham, O. *The Making of English Landscape*, Dent, 1987.

Rudwick, M.J.S. *The Great Devonian Controversy: The Shaping of Scientific Knowledge
Among Gentlemanly Specialists*, University of Chicago Press, 1985.

Rudwick, M.J.S. *The Meaning of Fossils: Episodes in the History of Palaeontology*,
Macdonald, 1972.

Sutcliffe, A.J. *In Search of Ice Age Mammals*, Natural History Museum Publications,
2nd ed., 1991.

Index

Abbotsford, 116, 117
Aberystwyth Grits, 95
Africa, 34, 38, 44, 45, 46, 100
agglomerates, 140, 270, 271–2, 295
Aldeburgh, 276, 277, 282, 283, 288
Alport Castle, 156
Alum Bay, 198, 249, 250, 251, 252, 256, 265
amber, 284
ammonites
  Dactylioceras, **178**
  Hildoceras, 188
  Parkinsonia, 180
  Titanites, 196–7
  disappearance of, 215, 250
  in Coal Measures, 150
  in Gault Clay, 210
  in Oxford Clay, 191–2
  in Portland Stone, 196–7
  near Lyme Regis, 177–8, 179–80
  near Whitby, 188
  mentioned, 3, 121, 129, 156, 162
Anglesey, 92, 126
anhydrite, 168
Anning, Mary, 175, 178
anticlines, 87, 199, 200, 295
Antrim, 268, 282
Archaean age, 34
architecture
  Chalklands, 223–7
  Dartmoor, 170
  East Anglia, 291–2
  Edinburgh, 140
  Exmoor, 125
  Highlands, 69–70
  imitation Jurassic, 187
  Lake District, 104
  Millstone Grit, 154
  Old Red Sandstone, 116, 119
  Oolitic limestone, 181–2, 184, 194–5
  Skye, 274
  slates in, 85–6, 122
  Southern Uplands, 116–17
  Wales, 10, 13, 96–7, 119
  Weald, 200–1, 210–11
  Yorkshire, 136–7, 190
Ardnamurchan, 72, 268, 271
Arkell, W.J., 181, 191
Arran, 70–2, 98, 113, 164, 268, 272, 273, 274
arthropods, 112–13, **113**, 115
Arthur's Seat, Edinburgh, 17, 140, 141
ash woodland, 140
Ashdown Forest, 208
Ashdowne House, 264
asphodel, bog (*Narthecium ossifragum*), 155
Atherfield Point, 212
Atlantic Ocean, 37, 38, 39, 40, 42, 43, 45, 122, 128, 161, 173, 267, 268, 273

Australia, 26, 34, 38, 46, 80, 95, 111
Avebury, 222–3, 264
Avon Gorge, 128
Axell, Herbert, 290

Bagshot Beds and Sands, 256, 263, 265
Bailey, E., 51, 63, 64
Balnakiel, 39, 48
Barra, **28**
barrows, 222
Barton Formation (Barton Clay), 253, 265
Barton on Sea, 252, 253, **254**
basalt, 17, 268–9, 270, 295
basement, 33–4, 35, 37
Bath, 3, 181, 182, 191, 202
batholiths, 168, 295
bed/bedding, 4–5, 18, 63, 110, 218, 295
beechwoods, 238–43
Beinn na Dubhaich, 272
Belvoir, Vale of, 148
Bembridge Marls, 265
Ben Arkle, 42
Ben Arnaboll, 52
Ben Lawers, 69
Ben Lui, **71**
Ben Nevis, 60, 61
Berwick Coast, 75, **77**
Billingham, 168
Billings, Elkanah, 40
Birdlip Hill, 184–5
birds, 97, 142, 156, 170, 193, 222, 290, 291, 292
Black Mountains, 117, 118, 119
Blackshore, 284
Blue Lias, 21, 177 *see also* Lias
Bodiam Castle, 210
Bodmin Moor, 17, 168, **169**, 171, 172
bogs, 67, 69, 155, 170
Bonchurch, 213
Bordes, M., 229, 230
Borrow, George, 96
Botallack mine, **171**
boulder clay, 188, 282, 295
Bovey Tracey, 266
Bowerman's Nose, 170
Bowland Forest, 153, 156
Box Hill, 208, 240, 243
box trees, 243
Boxstones, 267
brachiopods, 89, 98, 101, 121, 123, 167, 223, 235, 276
Bracklesham Group, 264, 265
Bradford-on-Avon, 184
Braes, 115
Breckland, 291
Brecon Beacons, 117, 118
Brendon Hills, 119, 123
Brent Tor, 186
bricks, 187, 191, 211–12, 225, 291–2
Bridport Sands, 180
Brimham Rocks, 156

Broadstairs, 218
Brodick Castle Gardens, 70–1
Brontë sisters, 153
Brownstones, 118, 119
Bullard, Sir Edward, 161
Burren, the, 128, 132, 139, 142

Cader Idris, 88, 90
Caer Caradoc, 83, 92, 94
Cairngorms, 14, 46, 60, 61, 62, 72, 169
Cairnsmore of Fleet, 75
Caithness, 67, 109, 110
Caithness Flags, 110–11
caldera, 270
Caledonian mountains
  erosion of, 45, 60, 107, 108, 109, 110, 112, 116, 117, 125, 126, 162
  formation of, 8, 11, 45, 107
  in Ireland, 74, 80
  in Lake District, 102, 127
  in Scotland, 46, 57–70, 72, 74, 75–6, 79
  in Shropshire, 104
  in Wales, 82, 85, 96, 97, 117
  see also names of mountains
Caledonian trend, 46, 57, 60, 72, 73, 74, 80, 84, 92
Callander, 57
Callaway, Charles, 50
Camborne School of Mines, 172
Cambrian period
  in Scotland, 35, 39, 42, **43**, 51, 54, 72
  in Wales, 12–13, 82, 83–8, 92, 99
  on geological maps, 23
Campsie Fells, 140
Canada, 30, 34, 41, 42, 70
Carboniferous limestone
  detailed discussion of, 126–40
  mentioned, 23, 46, 102, 141, 142–3, 152, 153, 154, 163, 215, 282
Carboniferous period
  and coal, 145, 146, 150
  limestone *see* Carboniferous limestone
  in South West, 156–9
  volcanic rocks, 140–1
  mentioned, 80, 113, 120, 122, 123, 151, 160, 162, 168–9, 209
Cardigan Bay, 95
Cardingmill Valley, 94
Carlops, 114
Carmarthen Bay, 128, 145
Carmyllie, 112
caves, 133–4, **135**
Cerne Abbas giant, 232
Chalk
  detailed discussion of, 215–48
  mentioned, 3, 23, 127, 138, 179, 198, 199, 202, 208, 209, 213, 257, 258, 261, 263, 264, 265, 275

Chapel Rock, 179
charcoal industry, 151
Charnia masoni, 94–5
Charnwood Forest, 94, 163
Cheddar Gorge, 128, 133, 202
Cheddar pink, 139
cheese, 166–7
Cheshire, 160, 165, 166 *see also* names of places
Cheshire Basin, 163, 165, 167, 168
Chesil Beach, 192, 193
Cheviot Hills, 116
Chilterns, 239, 243, 246, 247, 258, 263
china industry, 151–2, 172
Chippel Bay, 177
Church Stretton, 83, 92, 93, 94, 123
Cinque Ports, 201
citrine, 284
clay, 20, 21, 151, 176, 180, 181, 186, 190, 192, 194, 200, 208, 211, 212, 250, 253, 263, 291
  see also Barton Formation; Gault Clay, London Clay; Oxford Clay
clay-with-flints, 248, 295
cleavage, 85, 86, 122, 296
Cleddau Valley, 3, 4
clematis, wild, 243
Cleveland, 187
Cleveland Hills, 189
Clougha Pike, 156
Clovelly, 159
club mosses (*Lycopodium*), 68, 97, 149
clunch, 223
coal, 3, **4**, 127, 128, 141, 144–51, 152
  see also Coal Measures
Coal Measures, 127, 141, 145, 147, 148, 149, 150, 151, 152, 154, 156
Coalbrookdale, 151
coccoliths, 217
Collyweston 'slates', 183, **183**
Comley, 83, 93
conglomerates, 58, **59**, 74, 78, 82, 112, 264, 296
Conic Hill, 59, 60
Connemara, 60, 169
continental drift, 38, 46, 160–1
copper, 103, 171–2
coral, 3, 100–1, 121, 126, 129, 162, 193
Corallian limestone, 193
Coralline Crag, 276, 277
Corn Ddu, 118
Cornbrash, 21, 190
Corndon Hill, 94, 104
Cornish Heath (*Erica cornubiensis*), 123
Cornwall, 119, 120, 163, 169, **171**, 172 *see also* names of places

Cotham Marble, 176−7, **177**
Cotswold Stone, 181
Cotswolds, 176, 181, 182, 183−4
Crags, 267, 276−7, 283
cramp balls (*Daldinia concentrica*), 140
cranberries, 62
cranesbill, bloody (*Geranium sanguineum*), 140
Craven, 129, 132
Creswell Crags, 134, **164**
Cretaceous period, 48, 169, 179, 198, 199, 200, 205, 209, 210, 213, 215, 218, 220, 234, 250, 251, 261, 263, 291
Criffel, 75
crinoid (sea-lily), 129−30, 143, 235
Cromer, 277, 286, **289**
Cromer Forest Bed, 283
Cromer ridge, 282
cromlechs, 55
cross-bedding, 110, 296
Croucher, Ron, 205
crowberries, 62
Crystal Palace Park, 260
Cuillins, 17, 272
Culm, the, 120, 157, 159
Culzean Castle, **106**, 113−14
Cuthbert, St, 142
Cwm Idwal, 90
cycads, 189

Dalradian Series, 72, 268
Darby, Abraham, 151
Dartmoor, 14, 20, 27, 68, 121, 156, 168, 170, 172
Dartmouth Castle, **120**
Dashwood, Francis, 237
Delabole, 122, 123
'Delabole butterfly', 123
Derbyshire, 18, 128, 132, 134, 137, 156, 163
 see also names of places.
Devil's Beef Tub, **77**, 78, 185
Devil's Chair, 185
Devil's Chimney, Leckhampton, 185
Devil's Chimney, 185−6
Devil's Dyke, 186, **247**
Devil's Kitchen, 90, 186
Devil's Punchbowl, 186
Devon, 119, 120, 121, 122−3, 156, 157, 163, 209, 223 see also names of places
Devonian period, 38, 46, 47, 58, 73, 75, 108−125, 156, 168, 172, 213
Dingle Bay, 128
dinosaurs, 160, 183, 188, 197, 205−6, **206**, 215, 232, 250, 251, 260−1
Dobb's Linn, 76, 78
dolerite, 33, 91
dolomite, 39, 41, 44, 50, 54, 167, 296
Dorset, 12, 24, 175−6, 177, 179, 181, 192, 193, 194, 198, 232 see also names of places
Doubler Stones, 156, **157**
Dove Holes, **135**
Dover, 201, 218
Down, County, 74, 282
drift, 21
drumlins, 282
dry valleys, **247**, 247−8
Dudley, 101

Dulverton, 125
Dunwich, 288
Durness, 29, 39, 40, 41, 43, 44, 45, 48, 49, 55
Durness Limestone, 47, 48, 55
Dwarfie Stone, 110
Dyfed, 9−10, 97 see also names of places
dykes, 270, 273−4

Earth, formation of, 25−7
East Anglia, 220, 223, 224, 225, **226**, 227, 248, 267, 275−94
Easton Bavents, 287
Edinburgh, 17, 140
Eglwyseg Mountain, 128
Eigg, 268, 269, 270
Eocene, 257, 258, 259, 265, 270
erosion
 in Carboniferous Limestone, 134
 Dorset coast, 197, 198
 East Anglia, 275, 276, 287
 Scotland, 31, 33, 58, 59, 60, 61, 66, 107, 112
 South West, 122, 124, 169
 mentioned, 8, 18, 281
Eskdale, 102
'Estuarine' Series, 187−8
Europe, 38, 43, 44, 45, 46, 268, 281
evaporites, 167, 296
Exmoor, 68, 119, 123, 125, 156

facies, 121, 129, 296
Fairlight Cove, 200, 205
Farne Islands, 141, 142
felspar, 33, 35, 65, 170, 272
fens, 286
ferns, 68, 97, 98, 124, 139, 149, 213
fish, 111, 117−18 see also trout
flax, fairy (*Linum catharticum*), 241
Fleet, the, 193
flints, 216, 218, 220, 224−31, **231**, 235, 261, 291
flora see vegetation
flushwork, 227
Foinaven, 42
foliation, 62, 66, 296
foraminiferans, 217, 234, 257, 280, 296
Ford, Trevor, 94
Forest of Dean, 128, 147, 151
fossils
 Charnia masoni, 94−5
 in Chalk, 215−17, 220, 223, 232, 233, 235, 236
 in coal, 145−6, 149−50
 in Devon, 121, 123
 in East Anglia, 276, 277, 278, 283−4
 in Jurassic, 176, 177, **178**, 178, 183, 189, 191−2, **193**, 195−7
 in Magnesian Limestone, 167
 in Scotland, 39, 40, **41**, 61, 68, 74, 76, 78, 110−11, 112−13, 114, 115, 149−50
 in Shropshire, 83, 95, 101
 in Tertiary, 251−2, 253, 256, 257, 259, 264, 265, 266, 267, 271
 in Wales, 4, 5, 6, 7, 8, 9, 12, 81, 82, 84, 85, 87−8, 89, 90−1, 96, 98, 146
 in Weald, 205−7

tree, 149, 188, 189, 197, 264, 271
 mentioned, 3, 18, 34, 35, 39, 92, 100, 109, 110, 118, 156, 279, 280, **293**
 see also ammonites; brachiopods; graptolites; plant fossils; pollen; trilobites
foxgloves (*Digitalis*), 97
Frodingham Ironstone, 187
Fuller's Earth, 185
fungi, 69, 140, 241−2, 255

gabbro, 17, 270, 272, 296
Gargunnock Hills, 141
garnets, **32**, 64−5
Gault Clay, 179, 204, 209, 210, 211−12, 213
Geikie, Archibald, 48, 49, 50, 51, 116, 141
geological maps, 20−3, 51, 176, 299
geological time, **22**, 23−5
Giant's Causeway, 129, 268−9
Girvan, 78, 113
Gladiolus illyricus, 255
Glen Clova, 62
Glen Valtos, 31
globe flower (*Trollius europaeus*), 139
Globigerina Ooze, 234
Glyders, 88, 90, 98
gneiss, 29, 62, 65, 66, 296 see also Lewisian Gneiss
Goat Fell, 71, 72, 272
Gog Magog Hills, 223, 275
gold, 99, 100
Goredale Scar, 132
gorse, 208
Gower Peninsula, 128, 134, 281
granite
 definition of, 296
 Ireland, 61, 62
 Lake District, 102
 Scotland, 20, 33, 46, 60, 61, 62, 65, 72, 75
 South West, 168−71, 173, 174
 Tertiary, 272, 273
graptolites, 76, 78, **78**, 90−1, 95, 96, 103, 296
gravel, 256, 262, 263, 276, 282
Great Glen, **57**, 72−3, 92, 111
Great Scar, 132
Greenland, 26, 29, 40, 41, 42, 44, 118
Greensand, 186, 208, 209, 212, 213, 221
grits, 74, 75, 76, 83, 95, 96, 100, 122, 156, 157, 190, 296
 see also Millstone Grit
Gryphaea, 192, **193**
gypsum, 168

Hadrian's Wall, 142
Ham Hill Stone, 177
Hambleden, 246
Hampshire Basin, 197, 232, 252, 253, 257, 261
Hampstead Heath, 262
Hardraw Force, 137
Harlech Dome, 82, 83, **84**, 84
Harris, 29, 30, 31, 33
Harris, Tom, 188
Hartland Point, 159
Hastings Bed, 200
Haverfordwest, 3, 7, 8, 9, 11, 13
Haworth, 153

heath family (*Ericaceae*), 62, 105, 123, 155, 189
Hebrides (Western Isles), 29−33, 35, 267−74
 see also names of islands
Heddle, Matthew, 55
hedges, 209
Hercynian, 122, 123, 124, 125, 128, 146, 156, 157, 160, 162, 169, 174, 296
Herefordshire, 108, 117, 119, 120, 165
Herstmonceux Castle, 211
Hertfordshire 'Pudding Stone', 264, **265**
High Peak, 153, 155 see also Peak District
High Wycombe, 237, 263
Highland Boundary Fault, 57, 58, 59, 61, 70, 72, 74, 112, 155, 157
Highlands, 15, 19, 20, 27, 29, 34, 46, 48, 49, 50, 51, 57, 59−70, **71**, 72−3, 74, 79, 82, 97
Hog's Back, 222
holloways, 208−9
Holy Island, 127, 142−3, **143**
hornblende, 65, **66**
Horne, John, 40, 51
horsetail, 206
House, Michael, 179
Hunstanton, 223
Hutton, James, 75, 116

Iapetus Ocean, **39**, 45, 46, 47, 54, 60, 63, 76, 79, 80, 81, 82, 86, 90, 95, 97, 99, 100, 102, 107, 267
ice: detailed discussion of Ice Ages, 277−84; mentioned, 8, 20, 30, 31, 58, 60, 62, 66, 88, 89, 91, 92, 100, 102, 105, 107, 114, 118, 122, 134, 150, 169, 188, 189, 232, 248, 256, 272, 294
ichthyosaurs, 3, 178
igneous rocks
 distribution of, 20
 formation of, 17, 296
 see also volcanic rocks; names of rocks
Iltay Nappe, 72
induration, 264, 296
Ingleborough, 130, **131**
inliers, 93, 94, 95, 101, 113, 119, 128
intrusions, 61, 72, 75, 87, 169, 272, 297
Inverewe, 71
Inverness, 111
Ireland, 23, 37, 60, 61, 72, 74, 80, 99, 128, 132, 232, 268, 282 see also names of places
Irish Sea, 37, 46, 60, 74, 167
iron, 108, 151, 201, 204, 207
Ironbridge, 151, **152**
ironstones, 187, 193
Isle of Man, 37, 104
Isle of Wight, 185, 198, 200, 209, 210, 212−13, 232, **233**, 249, 265 see also names of places

Jedburgh, 116
jet, 192
Judd's Dykes, 72
Jurassic, 3, 23, 48, 104, 127, 160, 175−98

Keen, David, 256
Kennet and Avon Canal, 182
Kennet Valey, 261, 262
Kent, 146, 147, 201, 208, 211, 218, 231, 262
  see also names of places
Kentish Rag, 210, 211
Kilpeck Church, 119
Kimmeridge Clay, 181, 194
Kinder Scout, 153, 155, 156
Kingswear Castle, **120**
Knockan Cliff, 47, 48, 51, 52
Knole, 211
kyanite, 65
Kynance Cove, 123

Lake District, 37, 43−4, 45, 47, 54, 76, 90, 102−4, 127
Lambourn stream, 245−6
Lamont, Dr Archie, 114
Lancashire, 128, 147−8, 154
Land's End, 168, 171, 172−3, **173**
landscape marble, 176−7, **177**
Lapworth, Charles, 50, 51, 52, 54, 76, 78, 79
laurel, spurge (*Daphne laureola*), 239
Laurentia, 41, 43, 44, 45, 46, 50, 52, 54, 61, 76, 78, 79, 80, 82, 297
lava, 38, 90, 102, 140, 141, 267, 268, 269, 270
lavender, sea (*Limonium*), 290
Lawrence, D.H., 148
Laxford, 30, 33
lead, 24, 100
Lesmahagow, 113
Lewis, 29, 30, 31
Lewisian Gneiss, **28,** 29−35, 39, 42, 50, 51, 53
Lias, 12, 18, 21, 177, 178, 179, 180, 181, 185, 187, 188, 190, 192
lichens, 10, 97, 124, 139
limestone
  Carboniferous *see* Carboniferous Limestone
  Jurassic, 3, 176, 180−1, 182−3, **183,** 187, 190, 191, 193, 194, 197
  Magnesian, 167
  Oolitic *see* Oolites/Oolitic limestone
  pavements, **130,** 132, 139
  in Scotland, 39, 40, 41, 42, 47, 48
  mentioned, 18, 44, 50, 82, 101, 121, 123, 176, 202, 266
Lindisfarne *see* Holy Island
Little Moreton Hall, 166, **166**
liverworts, 213
Lizard, the, 123
Llanberis, 85, 86
Llandilo, 98, 99, 117
Llwyd, Edward, 98
Llyn Idwal, 88, 90, 98
Llyn-y-cwn, 90, 186
Loch Coruisk, 272
Loch Doon, 75
Loch Eil Psammite, **31**
Loch Eriboll, 41, 49, 50, 52
Loch Lomond, 58, 112
Loch Ness, 72−3
Loch Seaforth, 31
Loch Torridon, 34
Lochranza, 71

London, 144, 155, 182, 194, 199, 210, 212, 244, 258, 259−60, 262
London Basin, 232, 252, 258, 261, 262
London Clay, 256, 258, 259, 261, 262, 265, 277
Longmynd, 92, 93, 94, 123
Lulworth Cove, 193, 197, 198, 200
Lundy Island, 17
Lyme Bay, 120, 209
Lyme Regis, 175, 177, 179, 210
Lyn, River, 124−5

Maclean's Nose, 271−2
Magnesian Limestone, 167
maidenhair tree (*Gingko*), 188−9
Maitland, F.W., 164
Malham, **131,** 136
Malham Cove, 133
Malham Tarn, 133
Malvern Hills, 94
Mansfield Woodhouse, 148
marble, 18, 272
Marlborough Downs, 209, 221, 222, 248
Mason, Roger, 94, 95
Matlock, 137, 154
Menai Strait, 92
Mendip Hills, 128, 133, 134, 139, 163
Mesolithic industries, 230
Mesozoic, 162, 198, 235, 248, 297
metamorphic rocks
  formation of, 17, 18−20, 297
  vegetation associated with, 20
  see also names of rocks
mica, 19−20, 48, 62, 65, 66, 170
Midland Valley, 57, 74, 111, 113, 114, 117, 127−8, 141, 145, 147, 164
Midlands, 163, 165, 176 *see also* names of places
migmatites, 33, 297
milkwort (*Polygalum*), 241
Miller, Hugh, 111
Millstone Edge, 153
Millstone Grit, 127, 130, 138, 153, **154,** 154, 155, 156
Minack Theatre, 173−4
Minchin Hole, 134, 280
minerals, 32, 64−6, 99−100, 108, 121−2, 261
mines/mining, **71,** 99, 100, 103, 146−7, **171,** 171, 172
Minsmere, 290
Miocene, 267, 270, 276
Moffat, 76, 78, 185
Moine
  Schists, 48, 49, 50, 51, 72
  Thrust, 15, 51, 52, 53, 54, 55, 58, 59, 63, 76, 79, 120
Moon, 25, 26
Moor Grit, 189
moorland, 67−8, 84, 125, 155, 170, 189, 190
moraines, 282
Moray Firth, 111
morels, 242
Mountain Spiderwort (*Lloydia serotina*), 98
Much Wenlock, 101
mudstones, 7, 12, 62, 74, 87, 100, 297
Mull, 190, 268, 269, 270−1

Murchison, Sir Roderick, 48, 49, 50, 51, 52, 82
mylonite, 52, **52,** 53, 297
myrtle, bog, 255

Nantlle Slate belt, 84−5, 96
nappes, 63, 64, 157, 297
National Trust, 10, 208, 243
Neanderthal Man, 278
Needles, the, 232, **233,** 249, 250
Neolithic industries, 230
New Forest, 253−6
New Red Sandstone, 23, 108, 125, 156−7, 162−3, 164, 165
Newcastle-upon-Tyne, 147
Newfoundland, 40, 42, 44, 62
Nicol, James, 48, 49
Norfolk, 224, 225, 229, 275, 277, 280, 282, 283, 285, 292, 294
Norfolk Broads, 292, 294
North America, 38, 41, 42, 44, 51, 268
North Downs, 199, 221, 232, 239, 240, 258
North York Moors, 181, 189−90, 281
Northleach, 184
Norway, 45, 109
Norwich School, 285−6
Nottingham, 163, 164, 167
nummulites, 257, 258

oaks, 255
obsidian, 228−9
Ochil Hills, 115
Ockley, 205
oil, 194
Old Man of Hoy, 109
Old Red Sandstone, 22, 108−21, 123, 125, 126, 162, 165
Oolites/Oolitic limestone, 180, 181−2, 184, 185, 194, 297
Orcadian Basin, 109
orchids, 139, 239, 240
Ordovician
  Iapetus *see* Iapetus Ocean
  in Ireland, 80, 99
  in Lake District, 43−4, 102, 103, 104
  in Scotland, 40, 41, 61, 74, 76, 78
  in Shropshire, 94, 104, 105, 185
  in Wales, 44, 82, 88, 89, 90, 92, 95, 98, 99
  mentioned, 23, 47, 48, 100, 116
Orford Ness, 288
Orkney, 109, 110
Orwell, George, 144
Osmington Mills, 193
Owen, Sir Richard, 260
Oxford, 190, 191, **195,** 195
Oxford Clay, 21, 167, 181, 190−1, 192, 193, 210
Oxfordshire, 182, 184, 186

Palaeocene, 265
Palaeolithic period, 230, **231,** 293
Palaeozoic period, 23, 60, 160, 185, 297
Pangaea, 38, 39, 45, 47, 160, **161,** 161−2, 168
pantiles, 292
pargetting, 224, **226**
Peach, Ben, 40, 51
Peak District, 27, 128, 130, 134, 153, 155

Peak Fault, 188
peat, 30, 67, 115, 286, 287, **289,** 292
Pembrokeshire, 9, 10, 18
  see also names of places
Pen-y-Fan, 118
Pengelly quarry, Delabole, 122, 123
Pennines, 20, 127, 129, 130, 132, 133, 134, 137, 147, 153, 154, 167 *see also* names of places
pennywort (*Cotyledon*), 97
Penrhyn Quarry, Bethesda, 84, 85, **86,** 86
Pentland Hills, 114, 115
Permian age, 23, 42, 45, 108, 146, 160, 161, 162, 163, 164, 165, 167, 168, 209
Pewsey, Vale of, 209
Phragmites reed, 224, **289,** 290
Pickering, Vale of, 189, 281
Pinhay Bay, 176, 179
Pipe Rock, 42, **43,** 49
plant fossils, 115, 129−30, 143, 145−6, 188, 206, 213, 259
plate tectonics, 38, 90, 161
Plateau Beds, 118−19
Playfair, John, 75, 76
Pleistocene period, 62, 89, 91, 105, 169, 232, 256, 277, 279, 280, 281, 284, 286, 287, 291
plesiosaurs, 3, 178−9, 192, 233
Pliocene, 276, 287
plutonic rocks, 17
Plynlimon, 88
Polden Hills, 177
pollen, 67, 68, 236, 256
pollution, 144−5, 154
poppy
  Himalayan, 70−1
  Welsh, 98
Port Talbot, 148
Porth-yr-Ogof, 133
Porthcurno, 173
Porthmadoc, 81
Portland Bill, 176, 192−3
Portland Stone, 194−5, **195,** 196
potholes, 133
Precambrian age, 34, 35, 42, 47, 58, 62, 72, 83, 93, 94, 95, 105, 110, 123, 126, 297
Prescelli Hills, 91
Proterozoic age, 34, 35
psammites, **31,** 62−3
pterosaurs, 183, 233
Purbeck, Isle of, 176, 194, 197, 198, 200
Purbeck Beds, 198
Purbeck marble, 10, 197

Quantock Hills, 119, 123, 163
quarries, 83, 84−5, 86, 112, 134−5, 153, 154−5, 182, 194
quartz, 33, 65, 99, 153, 170, 297
quartz-dolerite, 141, 274
quartzites, **66,** 105, 264
Quaternary, 212, 258
quoits, 171

Rackham, Oliver, 27, 164, 236
radiometric dating, 24, 29, 33, 35, 133−4
Rafflesia, 240
Reading Beds, 3, 256, 261, 265
Red Crags, 276−7, 283
Red Hills, 272
Redcar, 177

Rhaetic rocks, 176
Rheidol gorge, 96
rhododendron, 263
Rhum, 35, 72, 268, 272
Rhynie Chert, 115
Ridgeway, The, 222, 235
Ringstead Bay, 193
Roach, 195
Rockall, 273
Roman Steps, 83
'root zone', 63
Ross and Cromarty, 35, 67
Rye, 200, 201, 211

St Austell, 172
St Bride's Bay, 87
St David's, 3, 9, 10, **11,** 13, 97, 121
St David's Head, 87
St David's Peninsula, 10–13, 82, 87, 88
St Kilda, 272–3
St Michael's Mount, **124,** 173, 186
St Michael' village, 264, **265**
Salisbury Cathedral, 197
Salisbury Plain, 91, 209, 222, 224, 232, 249
salt, 162, 167, 168
salt marshes, 290
samphire, 290
sandstones, 10, 18, 50, 51, 62–3, 75, 82, 87, 123, 129, 145, 152, 153, 155, 163–4, 179, 185, 187, 188, 189, 190, 200, 204, 210, 250, 262, 267, 277, 287 *see also* Greensand; New Red Sandstone; Old Red Sandstone; Pipe Rock; Torridonian rocks
sarsenstones, 223, 264, **266**
saxifrage, 97–8, 143
Scandinavia, 45, 62
Scarborough, 188
schist, 18, 48, 49, 50, 51, 52, 62, 66, **66,** 297
Schistose Grit, 72
Scilly Isles, 172
Scotland
  Caledonian, 57–73
  Carboniferous, 127–8, 140–1
  coal, 145, 147, 150
  and great divide, 39–43, 44, 45, 46, 47, 48–55
  Lewisian Gneiss, 29–34
  Old Red Sandstone, 109–17
  Southern Uplands, 74–80
  Tertiary, 267–74
  Torridonian, 34–5
  mentioned, 14, 15, 19, 20, 164, 281
Scots pine, 68
Scott, Sir Walter, 116–17, 142, 185
Scourie, 29–30, 33, 274
sea urchins, 216, 235
Sedgwick, Adam, 82
sedimentary rocks
  formation of 8, 17, 18, 19, 297
  vegetation associated with, 20
  see also names of rocks
Selbourne, 209, 247
serpentine, 123
Severn Valley, 14, 163
shales, 3, 10, 18, 62, 74, 76, 78, 80, 81, 82, 87, 91, 95, 96, 103, 123, 129, 145, 157, 176, 187, 197, 298

Shanklin, 212
Shap Granite, 282
Sheppey, Isle of, 259
Shropshire, 44, 50, 83, 88, 90, 92, 95, 97, 100, 101, 104–5, 117, 147, 151 *see also* names of places
Siccar Point, 75, 76, 116
Silbury Hill, 222, **236**
silica, 115, 218, 264, 298
silicon, 66
sillimanite, 65
siltstone, 118, 162
Silurian period
  in Scotland, 61, 74, 75, 76, 78, 114, 116
  in Shropshire, 101, 117
  in Wales, 3, 4, 5, 6, 9, 48, 82, 95, 96, 97, 99, 100
  mentioned, 23, 38, 94, 103, 108, 113, 117, 282
Skiddaw, 43, 102, 103
Skye, 17, 29, 35, 41, 48, 72, 190, 268, 269–70, 272, **273,** 274
slate, 10, 18, 75, 84, 85, 86, 96, 97, 122, 123
Smith, William, 21, 176, 182, 190
Smoo Cave, 39, 48, 55
Snailbeach, 100
Snowdon, 88, 89, 90
Snowdon lily, 98
Snowdonia, 82, 84, 89, 90, 97, 98, 186
Snuff Box Bed, 180
soil, 20, 21, 31, 61, 164, 209, 255–6, 262, 263
Solway Firth, 37, 46, 102, 163
Somerset, 119, 128 *see also* names of places
South Downs, 221, 232
South Ronaldsay, 109
South Wales coalfield, 117, 128, 134, 147, 148
Southern Uplands, 74–80, 95, 116, 185
Southern Uplands Fault, 74, 114
Southwold, 227, 278
spas, 202–4
speedwells, 291
Sphagnum moss, 69, 254
Spitsbergen, 40, 41, 42, 98
Spode, Josiah, 152
springs, 201, 202, 204, 245
springtails, 115, **115**
Spurred Coral Root (*Epipogium aphyllum*), 239–40
Staffa, 269
Staffordshire, 151, 152, 163, 165
stalagmite, 133–4, 279
steel, 151
Stiperstones Quartzite, 21, 105, 185
Stiperstones outcrop, 104–5
Stirling, 141
Stonehenge, 91–2, 222, 223, 264, **266**
Stonesfield 'slates', 183
Storr, The, 269–70
Stroud, 184, 185
Strumble Head, 17, 90–1
subduction, 79, 298
Suffolk, 224, 225, 227, 231, 275, 276, 277, 282, 283, 284, 286, 287, 291
Sugar Loaf, 119
Suilven, 34, 35, **37**

sundew (*Drosera*), 40, 68, 155
Sutherland, 42, 53, 67
Sutherland, Graham, 97
Swaledale, 130, 134, 137
sweet cicely (*Meum*), 138
Swinburne, Algernon, 213
syncline, 87, 117, 120, 157, 298

Tamar valley, 172
Tay Valley, 115
Teifi valley, 95
Tertiary period, 72, 190, 232, 249–74, 276
Test, River, 246, 247
Thames, River, 191, 258, 261, 280
Thames Valley, 3, 212, 262
thatch, 224
Thornton Force, **138**
Thurso, 110, 111
thyme, wild, 241
till, 282, 298
timber-framed buildings, 165–6, **166,** 223–4
tin mines, **171,** 171, 172
Tintagel, 119, 120
toothwort (*Lathraea squamaria*), 240
Torbay, 121
Torquay, 119, 121
Torridonian rocks, 35, **37,** 39, 42
tors, 170
Touch Hills, 141
trees, 67–8, 140, 188–9, 238, 243, 255
  and coal formation, 146, 149
  fossils, 149, 188, 189, 197, 264, 271
Trent, River, 163, 164
Triassic age, 39, 45, 108, 146, 160, 162, 163, 164, 165, 167, 168, 176, 209
trilobites
  agnostids, 87–8, **88**
  Angelina sedgwickii, 81, 83, 99, 123
  Calymene, **7,** 17, 101
  Conophrys salopiensis, 50
  Lloydolithus, 99
  Ogygiocarella debuchii, 98
  Olenellus, 42
  Paradoxides davidis, 12–13, 87
  Petigurus, 40–1, **41,** 42–3, 44, 45, 54, 61
  Phacops, 121
  Pseudatops viola, 86
  disappearce of, 160, 162
  in Devon, 121
  in Dudley, **7,** 101
  in Lake District, 44, 103
  in Scotland, 39, 40–1, **41,** 42, 61, 78
  in Shropshire, 50, 95
  in Wales, 3, 5, 6, 7, 8, 9, 12–13, 34, 44, 81, 82, 84, 86, 87–8, **88,** 98–9
trout, 137, 246–7
truffles, 242
tufa, 137
Tunbridge Wells, 199, 200, 201, 204, 210
turbidite rocks, 63, 64
Tyndrum lead mine, **71**

Uffington White Horse, 231–2
Uig Hills, 30, 33
Uist, 29, 30

unconformity, 75, **77,** 112, 116, 157, 209, 298

vegetation
  affected by geology, 20
  Arran, 70–1
  Carboniferous Limestone, 138–40
  Chalk, 238–45, 248
  East Anglia, 276, **289,** 290–1
  gravels and sands, 262, 263
  Greensand, 208, 213
  Highlands, 61–2, 67–9
  Millstone Grit, 155–6
  New Forest, 254–5
  Shropshire, 105
  South West, 123, 124, 170
  Wales, 97–9
  volcanic rocks, 17, 80, 89, 90, 95, 102, 115, 140–1, 268–70, 271

Wales, 3–13, 14, 17, 20, 34, 44, 68, 81–92, 95–100
  see also South Wales coalfield
Walker, William, 205
Walton, Izaak, 137
Warwickshire, 163, 165
Waun fach, 118
Weald, 27, 146, 151, 186, 199–213
Weald Clay, 204
Weald Stone, 210
Wedgwood, Josiah, 151–2
Welsh basin, 95, 96
Wenlock Edge, 94, 101
Wensleydale, 130, 134, 137
West Onny, River, 105
West Wycombe, 237
Westbury White Horse, **221**
Westray, 110
Westwood marshes, 290
Whin Sill, **141,** 141–2
Whitby, 176, 188
White, Gilbert, 247–8
White Horse, Vale of the, 221, 231
Whitecliff Bay, 252, 265
Whiting Ness, 112
Wigtownshire, 282
Wilson, J.T., 44, 46
Wiltshire, 239, 264 *see also* names of places
Winchelsea, 200, 201
Winterbourne stream, 246
Witney, 187
Wood, Stan, 149–50
wood avens (*Geum urbanum*), 213
Wookey Hole, 133, 134, 280
Worcestershire, 119, 165
Wordsworth, William, 53, 55
Wrekin, the, 94, 95
Wren's Nest, 101–2, **103**
Wye Valley Gorge, 128

Y Garth, 81, 83
Yes Tor, 170
yews, 243
Yoredale rocks, 136
York, Vale of, 146, 165
York Stone, 154–5
Yorkshire, 127, 136, 139, 154, 167, 181, 187–8, 189, 190, 232, 282 *see also* names of places

Zechstein Sea, 167